CARBON NANOTUBES

Theoretical Concepts and
Research Strategies for Engineers

CARBON NANOTUBES

Theoretical Concepts and Research Strategies for Engineers

A. K. Haghi, PhD, and Sabu Thomas, PhD

APPLE ACADEMIC PRESS

Apple Academic Press Inc. | Apple Academic Press Inc.
3333 Mistwell Crescent | 9 Spinnaker Way
Oakville, ON L6L 0A2 | Waretown, NJ 08758
Canada | USA

©2015 by Apple Academic Press, Inc.

First issued in paperback 2021

Exclusive worldwide distribution by CRC Press, a member of Taylor & Francis Group
No claim to original U.S. Government works

ISBN 13: 978-1-77463-365-6 (pbk)
ISBN 13: 978-1-77188-052-7 (hbk)

Library and Archives Canada Cataloguing in Publication

Haghi, A. K., author
Carbon nanotubes: theoretical concepts and research strategies for engineers / A.K. Haghi, PhD, and Sabu Thomas, PhD.

Includes bibliographical references and index.
ISBN 978-1-77188-052-7 (bound)
1. Carbon nanotubes. I. Thomas, Sabu, author II. Title.

TA418.9.N35H34 2015   620.1'15   C2014-908299-1

Library of Congress Cataloging-in-Publication Data

Carbon nanotubes: theoretical concepts and research strategies for engineers / A.K. Haghi, PhD, and Sabu Thomas, PhD.

pages cm
Includes bibliographical references and index.
ISBN 978-1-77188-052-7 (alk. paper)
1. Nanotubes. 2. Nanostructured materials. 3. Carbon. I. Thomas, Sabu. II. Title.

TA418.9.N35H335 2015   620.1'15--dc23   2014048263

Apple Academic Press also publishes its books in a variety of electronic formats. Some content that appears in print may not be available in electronic format. For information about Apple Academic Press products, visit our website at **www.appleacademicpress.com** and the CRC Press website at **www.crcpress.com**

CONTENTS

LIST OF ABBREVIATIONS

A	Attractive Segment
AC	Activated Carbon
ACF	Activated Carbons Fibers
AF	Attractive Force
AFM	Atomic Force Microscopy
AP	Asymmetric Packing Segment
APP	Asymmetric Packing Process
ASA	Adsorption Stochastic Algorithm
ATRP	Atom Transfer Radical Polymerization
BD	Brownian Dynamics
BET	Brunauer-Emmet-Teller Method
CHC	Cahn–Hilliard–Cook
CNT	Carbon Nano Tube
CO	Carbon Monoxide
CPMD	Car–Parrinello molecular dynamics
CRTs	Cathode Ray Tubes
CS	Chitosan
CV	Cyclic Voltammeter
CVD	Chemical Vapor Decomposition
CVI	Chemical Vapor Infiltration
D	Directional Segment
DA	Dubinin-Astakhov Model
dE	Exact Differential Energy
DF	Directional Force
DFT	Density Functional Theory
DOM	Dissolved Natural Organic Matter
DPD	Dissipative Particle Dynamics
dP_s	Differential Entropy for Macrosystems
DRS	Dubinin-Radushkevich-Stoeckly Model
DS	Dubinin-Stoeckly Model
DTA	Differential Thermal Analysis
ED	External Force-Induced Directional factor
EDLC	Electric Double-Layer Capacitor

EF-F	External Force-Specific Functional Segment
ERM	Effective Reinforcing Modulus
F-BU	Fabrication Building Unit
FE-SEM	Field-Emission Scanning Electron Microscopy
FED	Field Emission Displays
FEM	Finite Element Method
FET	Field Effect Transistors
FOM	Federation Object Model
GCMC	Grand Canonical Monte Carlo Simulation
GIC	Graphite Intercalation Compounds
H–T	Halpin–Tsai
HK	Horvath-Kawazoe Model
HLA	High Level Architecture
HOPG	High Pyrolitic Graphite
HREM	High-Resolution Electron Microscopy
HRTEM	High Resolution Transmission Electron Microscopy
IHK	Improved Horvath-Kawazoe Model
ISS	Interfacial Shear Strength
ITO	Indium Tin Oxide
JG	Janus Green B
LB	Lattice Boltzmann
LDFT	Local Version of Density Functional Theory
LJ	Lennard-Jones Potential Function
M–T	Mori–Tanaka
MB	Methylene Blue
MC	Monte Carlo
MD	Molecular Dynamics
MDS	Molecular Dynamics Simulation
MFC	Microbial Fuel Cells
MH	Multi-Scale Homogenization
MO	Metal Oxide
MSA	Molecular Self-Assembly
MSC	Molecular Sieving Carbon
MSD	Micropore Size Distribution
N-CE	Nano-Communication Element
N-ME	Nano-Mechanical Element
N-PE	Nano-Property Element

N-SE	Nano-Structural Element
ND	Nguyen and Do method
NLDFT	Non Local Density Functional Theory
OMT	Object Model Template
OOP	Object-Oriented Programming
OPLS-AA	Optimized Potential for Liquid Simulations All-Atom
PA	Polyamide
PALS	Positron Annihilation Lifetime Spectroscopy
PC	Polycarbonate
PSA	Pressure Swing Adsorption
PSD	Pore Size Distribution
PVA	Polyvinyl Acetate
PVP	Poly Vinyl Pyridine
R	Repulsive Segment
R-BU	Reactive Building Unit
RF	Repulsive Force
RISC	Reduced Instruction Set Computer
RTI	Runtime Infrastructure
RVC	Reticulated Vitreous Carbon
RVE	Representative Volume Element
SA-BU	Self-Assembly Building Unit
SAM	Self-Assembled Monolayer
SANS	Small-Angle Neutron Scattering
SAXS	Small-Angle X-Ray Scattering
SOCs	Synthetic Organic Chemicals
SOM	Simulation Object Model
STM	Scanning Tunneling Microscopy
SUSHI	Simulation Utilities for Soft and Hard Interfaces
T&O	Taste and Odor
TDGL	Time-Dependent Ginsburg–Landau
TEM	Transmission Electron Microscopy
TEP	Thermoelectric Power
TETA	Triethylene Tetramine
TGA	Thermo-Gravimetric Analysis
Th	Thionine
TSA	Temperature Swing Adsorption
TVFM	Theory of Volume Filling of Micro pores

UFF	Universal Force Field
VLS	Vapor-Liquid-Solid
VSM	Vibrating Sample Magnetometer
XPS	X-Ray Photoelectron Spectroscopy
XRD	X-Ray Diffraction

LIST OF SYMBOLS

$\bar{\sigma}$	stress
v_f	volume fraction of filler
$\Pi_1 \& \Pi_2$	characterize the strength of surface forces field
A_{pot}	free energy of adsorption in Dubinin–Astakhov isotherm equation
$(L - d_a)_{p,max}$	pore width related to maximum value of function
ΔG_m	transformation from austenite to ferrite
A_{HK}	constant of HK method
a_m	monolayer capacity
$A_{max}, \bar{A} \& \sigma_A$	proportional to the parameter
a_{mi}	equilibrium amount adsorbed in micro pores
C_1, C_2, C_3, C_4	constants for adsorbate-adsorbent system
C_{BET}	BET constant on flat surface
c_s	complete filling concentration
d_a	diameter of the adsorbate molecule
d_A	diameter of the adsorbent atom
D_c^γ	diffusion constant of C in austenite
$D_{in} \& \chi(D_{in})$	differential PSD function
d_γ	austenite grain size after reheating of slab
$\frac{dA}{dx}$	derivative of adsorption potential
f_d	downstream fugacities
f_u	upstream fugacities
$\dot{G}$	grow rate

I_s	nucleation rate at unit area
$L_{av,I}$	average effective width of primary porous structure
$L_{av,II}$	average effective width of secondary porous structure
$L_{av,tot}$	average effective width of primary and secondary porous structure
L_{av}	average pore width
$\frac{L-d_a}{d_A}$	reduced effective pore width
$(L-d_a)$	effective pore width
n_0	number of austenite grains at a unit volume
N_{DRS}	values of adsorption
N_{mDRS}	values of maximum adsorption
$\dot{N}$	nucleation rate for unit volume
p_0	standard state pressure
P_e	permeability of membrane
p_s	saturation pressure of an adsorbate
q^{diff}	differential heat of adsorption
$S_{c,\alpha}$	total specific surface area calculate by high-resolution α_s-plot
S_{DFT}	surface of pores calculated for all pores
S_g	molar entropy of gas
V_{HK}	micropore volumes determined from HK
V_m	molar volume
V_m	molar volume of the adsorbent
W_0	Dubinin–Radushkevich isotherm equation
$W_{exp,tot}$	maximum value of the experimental adsorption
X_{dyn}	fraction dynamically recrystallized
X_e, Y_e	extended volume fraction
$X_{mi}(A)$	potential distribution in micro pores
$z_{max}, \bar{z}\&\sigma_z$	structural heterogeneity of microporous solids
γ_∞	surface tension of bulk fluid
γ_{me}	amount adsorbed at relative pressure per unit area of mesopore surface
γ_s	unit area of nonporous reference adsorbent surface
ε_0	adsorption characterized energy of standard state
$\lambda_1\&\lambda_2$	range of structural forces action
ρ_b^*	reduced number density of bulk fluid

v	velocity
ψ	wave function
δQ_{in}	inexact differential amounts of heat
δW_{in}	inexact differential amounts of work
Δx	spatial extension
ΔG^{ads}	free energy of adsorption equal to $RT\ \ln(p/p_s)$
ΔH^{ads}	enthalpy of adsorption
ΔH^{vap}	enthalpy of vaporization
ΔS^{ads}	entropy of adsorption
A	adsorption potential
a	total amount adsorbed
A, B	planes
a(k)	distribution of wave numbers
a^0_{mes}	monolayer capacity of mesopore surface
a^0_{mi}	maximum amount adsorbed in micro pores
a^0_s	monolayer capacity of reference adsorbent from standard adsorption isotherm
a_{mes}	amount adsorbed on mesopore surface
$a_{mes}(p/p_0)$	amount adsorbed on mesopore surface in relative pressure
a_{mic}	amount adsorbed on micro pore surface
$a_{mic}(p/p_0)$	amount adsorbed on micro pore surface in relative pressure
$a_s(0.4)$	amount adsorbed on surface of reference solid at pressure 0.4
$a_s(p/p_0)$	amount adsorbed on surface of reference adsorbent
b	an equilibrium adsorption constant
B	temperature-independent structural parameter of micropore sizes
c	average interstitial concentration
C	stiffness tensor
Ch	chiral vector
D	diffusion coefficient of C in austenite
d	tube diameter
D_0	constant Number
d^3r	volume element
D_{max}, D_{min}	minimum and Maximum pore size in kernel of NLDFT
dP	probability
E_0	characteristic energy of adsorption for reference vapor

erf(W)	error function of w parameter
$\varepsilon(x)$	non-uniform in polymer
F	conversion factor for N_2 adsorption
f	fluid fugacity
$f(H)$	pore size distribution function
$f(z)$	intrinsic molecular Helmholtz free energy of adsorbate phase
$F[\rho(r)]$	intrinsic Helmholtz free energy function
F(B)	Gaussian distribution function of the structural parameter B
F(L)	PSD of the heterogeneous solid adsorbent
F(Z)	distribution function characterizing heterogeneity of microporous structure
ω	frequency
h_{cr}	critical capillary film thickness
H_{max}	width of largest pore
H_{min}	width of smallest pore
I	intensity
J	local molar flux
$J(L - d_a)$	effective pore size distribution
J(X)	micro pore size distribution
k	structural parameter
k	wave vector
L	micropore width
L/d_A	reduced pore width
M	mass
M	mobility of interface
m	proportionality constant
$m = (\beta\kappa)^{-2}$	proportional coefficient
mic	calculated only for the range of micro pores
N_A	Avogadro number
p/p_o	relative pressure
q_1-q_1	difference heat of condensation and adsorption
r	average size of precipitates
R	universal gas constant
$r(p/p_0)$	pore radius
r/R	ratio of distance to radius

r_{cr}	critical capillary radius
S	entropy for nano/small systems
S_{BET}	surface area of BET method
S_{meso}	total surface area of mesopores
S_{mi}	total surface area of micropores
S_t	slope of low pressure of α_s-plot
suffix i	all elements in the system
T	time
t-plot	T-curve method
$t\left(\frac{p}{p_0}\right)$	statistical thickness of the film
V(A)	characteristic adsorption curve
V(r)	interaction potential of the fluid molecule
V_{meso}	mesopore volume
V_{micro}	micropore volume
V_t	maximum volume adsorbed
V_t-v	unoccupied pore volume
W^0_{mi}	limiting volume of adsorption
λ	wavelength
χ	fractions of elements
x	pore mole fraction in binary systems
X, Y	actual volume fraction
X(A)	adsorption potential distribution function
$X^*(A)$	non- normalized adsorption potential distribution function
x_0	micropore half width at the maximum of adsorption curve
$X_{mi}(A)$	micropore adsorption potential distribution function
y	bulk mole fraction in binary systems
Z	quantity associated with the micropore size
α	ferrite
β	similarity coefficient
γ	Austenite
Γ	Euler gamma function
δ	displacement of surface
δ, Δ	dispersion
ΔH	differential enthalpy

ΔS	differential entropy
ε	adsorption characterized energy
ε_{FC}	Len–Jones interaction parameter
ζ	constant equal to $\left(\dfrac{1}{\kappa^n}\right)$
η	Slope of linear segment of the α_s-plot
θ	chiral angle
Θ	degree of micropore filling
$\theta(L, P)$	local adsorption isotherm (kernel)
$\theta = a/a^0_{mi}$	relative adsorption
θ_{mes} (p/p_0)	relative surface coverage of mesoporous reference adsorbent
θ_{mi} (p/p_0)	relative surface coverage of micro porous reference adsorbent
$\theta_{mic}(z,A)$	local isotherm in uniform micropores
θ_s (p/p_0)	relative surface coverage of nonporous reference adsorbent
μ	chemical potential
ν	liquid molar volume
ν	parameter for the gamma distribution function
ρ	parameter of gamma distribution function
$\rho(z)$	local density of the adsorbed fluid
$\rho(P, H)$	density of N_2 at pressure P in a pore of width H
ς	proportionality constant
σ_{FC}	Len–Jones interaction parameter
τ	time when the new phase nucleates at plane B
$\Phi(r)$	external potential function
φij (r)	pair potential between 2atoms
$\chi(D_{in})$	normalized differential PSD function
$\Omega[\rho(z)]$	grand potential function
$\omega, '\Omega$	cross-sectional area

PREFACE

Carbon nanotubes, with their extraordinary mechanical and unique electronic properties, have garnered much attention in the recent years. With a broad range of potential applications including nanoelectronics, composites, chemical sensors, biosensors, microscopy, nanoelectro mechanical systems, and many more, the scientific community is more motivated than ever to move beyond basic properties and explore the real issues associated with carbon nanotubes-based applications.

Carbon nanotubes are exceptionally interesting from a fundamental research point of view. They open up new perspectives for various applications, such as nano-transistors in circuits, field-emission displays, artificial muscles, or added reinforcements in alloys. This text is an introduction to the physical concepts needed for investigating carbon nanotubes and other one-dimensional solid-state systems. Written for a wide scientific readership, each chapter consists of an instructive approach to the topic and sustainable ideas for solutions.

A large part of the research currently being conducted in the fields of materials science and engineering mechanics is devoted to carbon nanotubes and their applications. In this process, modeling is a very attractive investigation tool due to the difficulties in manufacturing and testing of nanomaterials. Continuum modeling offers significant advantages over atomistic modeling. Furthermore, the lack of accuracy in continuum methods can be overtaken by incorporating input data either from experiments or atomistic methods. This book reviews the recent progress in modeling of carbon nanotubes and their composites. The advantages and disadvantages of different methods are discussed. The ability of continuum methods to bridge different scales is emphasized. Recommendations for future research are given by focusing on what each method have to learn from the nano-scale. The scope of the book is to provide current knowledge aiming to support researchers entering the scientific area of carbon nanotubes to choose the appropriate modeling tool for accomplishing their study and place their efforts to further improve continuum methods.

CHAPTER 1

BASIC CONCEPTS

CONTENTS

1.1   INTRODUCTION

In this section, we focus on the pore size controlling in carbon based nano adsorbent to apply simulation and modeling methods and describe the recent activities about it. Significant progress has been made in this process throughout the recent years essential emphasis is put on the controlling and applications of both micro and mesoporosity carbons. Many novel methods are proposed such as catalytic activation, polymer blend, organic gel and template carbonization for the control of mesopore, and analyzing micro pore distribution in activated carbons assuming an array of semi-infinite, rigid slits of distributed width whose walls are modeled as energetically uniform graphite. Various kinds of pores in solid materials are classified according to the origin of the pores and the structural factors of the pores are discussed as well as the methods for evaluation of the pore size distribution with molecular adsorption (molecular resolution porosimetry), small angle X-ray scattering, mercury porosimetry, nuclear magnetic resonance, and thermo porosimetry. The main aim in the controlling of micro pores is to produce molecular sieving carbon (MSC), which applied in special membranes or produce some carbon composites which remove special contaminate from aqueous environments as adsorbent or EDLCs in electrochemical application or aero gel carbon structures. Despite difference in particle size, the adsorption properties of activated carbon and carbon based nano adsorbents (ACF-ACNF-CNF-CNT and CNT-Composites) are basically the same because the characteristics of activated carbon (pore size distribution, internal surface area and surface chemistry) controlling the equilibrium aspects of adsorption are independent of particle. The excellently regular structures of CNTs and its composites facilitate accurate simulation of CNT behavior by applying a variety of mathematical and numerical methods. The most important of this models include: Algorithms and simulation, such as Grand Canonical Monte Carlo (GCMC) simulation, Car-Parrinello molecular dynamics (CPMD), molecular dynamics simulation (MDS), LJ potential, DFT, HK, IHK, BJH, DA, DRS and BET Methods, ASA, Verlet and SHN algorithm and etc. Briefly, (in authors Ph.D. case study) this following research work includes:

- Preparing web of CNT-Textile composite (carbon section in an appropriate furnace and other section by electro spinning along CNT added in spinning drop).

- Stabilization process is performs in adequate temperatures in an appropriate furnace.
- Carbonization process is performs in adequate temperatures in an appropriate furnace.
- Activation process is performs in adequate temperatures in an appropriate furnace.
- Adsorption isotherms and characterizing of this composite will be determined and fitted.
- Adsorption parameters like PSD, S_{BET}, V_{meso} and V_{mic} for CNT-Textile composite and Textile section (with more probability, ACNF), calculated by BET, DFT, DRS and HK models and results are compared.
- Pore size distributions and PSD curves are determined from experimental isotherms like t-plot.
- Samples simulated using GCMC or MDS by assumption a adequate simulation cell (cylindrical slit-shape pore).
- Finally experimental and model results of ACNF and CNT-ACNF and its effect on PSD and selective adsorption are compared, adapted and model is verified.

1.2 ACTIVATED CARBON

Activated carbon (AC) has been most effective adsorbent for the removal of a wide range of contaminants from aqueous or gaseous environment. It is a widely used adsorbent in the treatment of wastewaters due to its exceptionally high surface areas which range from 500 to 1500 $m^2\,g^{-1}$, well developed internal micro porosity structure [1]. While the effectiveness of ACs to act as adsorbents for a wide range of pollutant materials is well noted and more research on AC modification are presented due to the need to enable ACs to develop affinity for special contaminants removal from wastewater [2]. It is, therefore, essential to understand the various important factors that influence the adsorption capacity of AC due to their modification so that it can be tailored to their specific physical and chemical attributes to enhance their affinities pollutant materials. These factors include specific surface area, pore size distribution, pore volume and presence of surface functional groups. Generally, the adsorption capacity increases with specific surface area due to the availability of adsorption site while pore size, and micro pore distribution are closely related to the composition of the AC, the type of raw material used and the degree of activation during production stage [3]. Here we summarize the various AC modification techniques and their effects on adsorption of chemical species from aqueous solutions. Modifications of AC in granular or powdered form were reviewed. Based on extensive literature reviews, the authors have categorized the techniques into three broad groups, namely, modification of chemical, physical and biological characteristics, which are further, subdivided into their pertinent treatment techniques. Table 1.1 lists and compares the advantages and disadvantages of existing modification techniques with regards to technical aspects, which are further elucidated in the following sections [1–4].

While these characteristics are reviewed separately as reflected by numerous AC modification research, it should be noted that there were also research with the direct intention of significantly modifying two or more characteristic and that the techniques reviewed are not intended to be exhaustive. The adsorption capacity depends on the accessibility of the organic molecules to the micro porosity, which is dependent on their size [5]. Activated carbon can be used for removing taste and odor (T&O) compounds, synthetic organic chemicals (SOCs), and dissolved natural organic matter (DOM) from water. PAC typically has a diameter less than 0.15 mm, and can be applied at various locations in a treatment system. GAC, with diameters ranging from 0.5 to 2.5 mm, is employed in fixed-bed adsorbents such as granular media filters or post filters. Each of these factors must be properly evaluated in determining the use of activated carbon in a practical application. The primary treatment objective of activated carbon adsorption in a particular water treatment plant determines the process design and operation; multiple objectives cannot, in most cases, be simultaneously optimized [5, 35].

It is critical in either case to understand and evaluate the adsorption interactions in the context of drinking water treatment systems. Activated carbons are prepared from different precursors and used in a wide range of industries. Their preparation, structure and applications were reviewed in different books and reviews. High BET surface area and light weights are the main advantages of activated carbons. Usually activated carbons have a wide range of pore sizes from micro pores to macro pores, which shows a marked contrast to the definite pore size of zeolites [41–46] (Table 1.1).

TABLE 1.1 Technical Advantages and Disadvantages of Existing Modification Techniques

Modification	Treatment	Advantages	Disadvantages
Chemical characteristics	Acidic	Increases acidic functional groups on AC surface. Enhances chelation ability with metal species	May decrease BET surface area and pore volume, May give off undesired SO_2 (treatment with H_2SO_4) or NO_2 (treatment with HNO_3) gases, has adverse effect on uptake of organics
	Basic	Enhances uptake of organics	May, in some cases, decrease the uptake of metal ions
	Impregnation of foreign material	Enhances in-built catalytic oxidation capability	May decrease BET surface area and pore volume
Physical characteristics	Heat	Increases BET surface area and pore volume	Decreases oxygen surface functional groups
Biological characteristics	Bio adsorption	Prolongs AC bed life by rapid oxidation of organics by bacteria before the material can occupy adsorption sites	Thick biofilm encapsulating AC may impede diffusion of adsorbate species

Previous researches showed that one of the most effective approaches for increasing mesopore volume of activated carbon is to catalyze the steam activation process by using transition metals or rare earth metal compounds, which can promisingly promote mesopore formation. The mechanism of mesoporous development is that the activation reaction takes place in the vicinity of metal/oxide particles, leading to the formation of mesopore by pitting holes into the carbon matrix. However, transition metals or rare earth metal compounds are still expensive and can seldom be reused on large scale, which limits industrial application of these processes, KOH plays three roles in the preparation of the activated carbons. First of all, it guarantees that the carbonization of coal is in solid phase. Besides, KOH can react with coal to form microspores, which provide "activation path" for steam, which can expand the microspore.

However, if there are too much microspores, combination of microspores will happen, resulting in the formation of much macro pore, and then can increase the ignition loss, so there must be an optimal addition of KOH in the precursor. Furthermore, KOH catalyzes the steam activation process. Because of high catalytic activity of KOH, acid washing process is used after carbonization process and before steam activation process to vary the content of K-containing compounds left in the char, and then the degree of steam activation can probably be varied. Consequently, regulation of the pore size distribution of activated carbon from coal is performed in two ways: the first is to change the addition of KOH in the precursor to produce optimal microspore in the char, and the second is to use acid washing process to vary the degree of catalytic effect of K-containing compounds in steam activation. The pore size distributions of coal-based activated carbons were successfully regulated by modifying the steam activation method the principles for the regulation of pore size distribution in the activated carbons were also discussed in this article, and we found that regulation of the pore size distributions of the activated carbons from coal are performed in two ways: the first is to change the addition of KOH in the precursor to produce optimal microspore in the char, and the second is to use the acid washing process to vary the degree of catalytic effect of K-containing compounds in steam activation [35–40, 177].

1.3   ACTIVATED CARBON FIBERS (ACF)

Activated carbons fibers have been prepared recently and developed a new field of applications. They have a number of advantages over granular activated carbons. The principal merit to prepare activated carbon in fibrous morphology is its particular pore structure and a large physical surface area. Granular activated carbons have different sizes of pores (macrospores, mesopore and microspores), whereas ACFs have mostly microspores on their surfaces. In granular activated carbons, adsorbents always have to reach microspores by passing through macro pores and mesopore, whereas in ACFs they can directly reach most microspores because microspores are open to the outer surface and hence, exposed directly to adsorbents. Therefore, the adsorption rate, as well as the amount of adsorption, of gases into ACFs is much higher than those into granular activated carbons [1–3].

In the recent work, the amount of adsorption of toluene molecules is much higher, and desorption proceeds faster in ACFs than granular activated carbons, effective elimination of SO from exhausted gases by using ACFs was too. A very high specific surface area up to 2500 $m^2 g^{-1}$ and a high microspore volume up to 1.6 $cm^3 g^{-1}$ can be obtained in isotropic-pitch-based carbon fibers For the preparation of these carbon fibers with a very high surface area such as 2500 $m^2 g^{-1}$, precursors which give a carbon with poor crystallinity are recommended; thus, mesophase-pitch-based carbon fibers did not give a high surface area, whereas isotropic pitch based carbon fibers did. Other advantage of ACFs is the possibility to prepare woven clothes and nonwoven mats, which developed new applications in small purification systems for water treatment and also as a deodorant in refrigerators in houses, recently reported. In order to give the fibers an antibacterial function and to increase their deodorant function, some trials on supporting minute particles of different metals, such as Ag, Cu and Mn, were performed. Table 1.2 presented Comparison between properties of activated carbon fibers and granular activated carbons [13, 46].

Activated carbon fibers, due to their micro porosity, are an excellent material for a fundamental study of H_2 adsorption capacity and enthalpy. Synthesized from polymeric carbon precursors, ACFs contain narrow and uniform pore size distributions with widths on the order of 1 nm. Images of ACFs from scanning tunneling microscopy have revealed networks of elongated slit-shaped and ellipsoid shaped pores. Edge terminations in graphitic layers are thought to be the most reactive sites during the steam/carbon dioxide activation process, resulting in a gradual lengthening of slit-shaped pores as a function of burn-off ACFs subjected to less burn-off will have smaller pore volumes and a greater abundance of narrow pores widths. With longer activation times, the pore volume increases and the pores grow wider. This offers a convenient control for an experimental study of the correlation between pore structure and hydrogen adsorption. In the current study, the pore size distribution (PSD) of activated carbon fiber is used to interpret the enthalpy and the capacity of supercritical H_2 adsorption [2, 175].

TABLE 1.2   Comparison Between Some Properties of ACF and GAC

	Activated carbon fibers	Granular activated carbons
	700–2500	900–1200
Surface area	0.2–2.0	~0.001
Mean diameter of pores (nm)	<40	From micro to macrospores

1.4   MOLECULAR SIEVING CARBONS (MSCS)

MSCs have a smaller pore size with a sharper distribution in the range of microspores in comparison with other activated carbons for gas and liquid phase adsorb ate. They have been used for adsorbing and eliminating pollutant samples with a very low concentration (ethylene gas adsorption to keep fruits and vegetables fresh, filtering of hazardous gases in power plants, etc.) An important application of these MSCs was developed in gas separation systems [1, 2]. The adsorption rate of gas molecules, such as nitrogen, oxygen, hydrogen and ethylene, depends strongly on the pore size of the MSC; the adsorption rate of a gas becomes slower for the MSC with the smaller pore size. The temperature also governs the rate of adsorption of a gas because of activated diffusion of adsorbate molecules in microspores: the higher the temperature, the faster the adsorption [47–49].

By controlling (swinging) these parameters, temperature and pressure of adsorb ate gas, gas separation can be performed. Depending on which parameter is controlled, swing adsorption method is classified into two modes, temperature swing adsorption (TSA) and pressure swing adsorption (PSA). Adsorption of oxygen into the MSC completes within 5 min, but nitrogen is adsorbed very slowly, less than 10% of equilibrium adsorption even after 15 min. From the column of MSC, therefore, nitrogen rich gas comes out on the adsorption process, and oxygen-rich gas is obtained on the desorption process. By using more than two columns of MSC and repeating these adsorption/desorption processes, nitrogen gas is isolated from oxygen. This swing adsorption method for gas separation has advantages such as low energy cost, room temperature operation, and compact equipment [50–53].

1.5   SUPER CAPACITORS (EDLCS) TO CONTROL PORE SIZE IN POROUS CARBONS

The main objective of this section is to provide a brief review of the pore size control that is an important factor influencing application of activated carbon or carbon nano structures adsorption in drinking water treatment or other adsorbent application. Different pore sizes in carbon materials are required for their applications. Therefore, the PSD in carbon materials has to be controlled during their preparation, by selecting the precursor, process and condition of carbonization, and also those of activation. A wide range broad distribution in pore size and shape is usually obtained in carbon materials. The control of pore size in carbons is essential in order to compete in adsorption performance with porous inorganic materials such as silica gels and zeolites, and to use the advantages of carbon materials such as high chemical stability, high temperature resistance and low weight. For applications in modern technology fields, not only high surface area and large pore volume but also a sharp pore size distribution at a definite size and control of surface nature of pore walls are strongly required. In order to control the pore structure in carbon materials, studies on the selection of precursors and preparation conditions have been extensively carried out and certain successes have been achieved [1–3]. Pore sizes and their distributions in adsorbents have to comply with requirements from different applications. Thus, relatively small pores are needed for gas adsorption and relatively large pores for liquid adsorption, and a very narrow PSD is required for molecular sieving applications. Macrospores in carbon materials were found to be effective for sorption of viscous heavy oils. Recent novel techniques to control pore structure in carbon materials can be expected to contribute to overcome this limitation [41–46].

One of newest application that shows importance of pore structure control in carbon materials is an EDLCor super capacitor that is an energy storage device that uses the electric double layer formed at the interface between an electrode and the electrolyte. EDLCs are well documented to exhibit significantly higher specific powers and longer cycle lifetimes compared with those of most of rechargeable batteries, including lead acid, Ni-MH, and Li-ion batteries. Hence, EDLCs have attracted considerable interest, given the ever-increasing demands of electric vehicles, portable electronic devices, and power sources for memory backup. The capacitance of an EDLC depends on the surface area of the electrode materials.

Therefore, activated carbons are necessary materials for EDLC electrodes because of their large surface area, highly porous structure, good adsorption properties, and high electrical conductivity. The electrochemical performance of EDLCs is related to the surface area, the pore structure, and the surface chemistry of the porous carbon. Various types of porous carbon have been widely studied for use as electrode materials for EDLCs. Their unusual structural and electronic properties make the carbon nanostructures applicable in, inter alia, the electrode materials of EDLCs and batteries. Activated carbon nano fibers are expected to be more useful than spherical activated carbon in allowing the relationship between pore structure and electrochemical properties to be investigated to prepare the polarizable electrodes for experimental EDLCs, EDLCs are well documented to exhibit significantly higher specific powers and longer cycle lifetimes compared with those of most of rechargeable batteries, including lead acid, Ni-MH, and Li-ion batteries [20, 34, 45].

Hence, EDLCs have attracted considerable interest, given the ever-increasing demands of electric vehicles, portable electronic devices, and power sources for memory backup. The capacitance of an EDLC depends on the surface area of the electrode materials. Therefore, activated carbons are necessary materials for EDLC electrodes because of their large surface area, highly porous structure, good adsorption properties, and high electrical conductivity. The electrochemical performance of EDLCs is related to the surface area, the pore structure, and the surface chemistry of the porous carbon. Various types of porous carbon have been widely studied for use as electrode materials for EDLCs. Their unusual structural and electronic properties make the carbon nanostructures applicable in the electrode materials of EDLCs and batteries. The principle of electrochemical capacitors, physical adsorption/desorption of electrolyte ions in solution, was applied for water purification by using different carbon materials [108–113].

This chapter is concerned with such pore control methods proposed by researchers that their ultimate goal is to establish a method with tailoring carbon material pore structures to reach any kind of application. Researchers would like too much effort that has made to control micro and mesopore in carbon materials, and prepare them in achieving the final goal. The presence of mesopore in electrodes based on CNTs, due to the central canal and entanglement enables easy access of ions from electrolyte. For electrodes built from multiwalled carbon nanotubes (MWCNTs), specific capacitance in a range of 4–135F/g was found. For single walled carbon nanotubes (SWCNTs) a maximum specific capacitance of 180F/g is reported. A comparative investigation of the specific capacitance achieved with CNTs and activated carbon material reveals the fact activated carbon material exhibited significantly higher capacitance. Super capacitor CNTs-based electrodes were fabricated by direct synthesis of nanotubes on the bulk Ni substrates, by means of plasma enhanced chemical vapor deposition of methane and hydrogen. The specific capacitance of electrodes with such nanotubes was of 49F/g. MWCNTs were electrochemically oxidized and their performance in EDLCs was studied [45–64–68].

1.6   CARBON NANOTUBE (CNT)

Researchers showed that carbon nanotubes are formed during arc-discharge synthesis of C_{60}, and other fullerenes also triggered an outburst of the interest in carbon nano fibers and nanotubes. These nanotubes may be even single walled, whereas low-temperature, catalytically grown tubes are multi walled. It has been realized that the fullerene-type materials and the carbon nano fibers known from catalysis are relatives and this broadens the scope of knowledge and of applications. This paper describes the issues around application and production of carbon nanostructures. Electro spinning is a simple and versatile method for generating ultrathin fibers from a rich variety of materials that include polymers, nano composites and ceramics. In a typical process, an electrical potential is applied between a droplet of a polymer solution, or melt, held at the end of a capillary tube and a grounded target. When the applied electric field overcomes the surface tension of the droplet, a charged jet of polymer solution is ejected. The following parameters and processing variables affect the electro spinning process: (i) system parameters such as molecular weight, molecular weight distribution and architecture (branched, linear, etc.) of the polymer, and polymer solution properties (viscosity, conductivity, dielectric constant, and surface tension, charge carried by the spinning jet) and (ii) process parameters such as electric potential, flow rate and concentration, distance between the capillary and collection screen, ambient parameters (temperature, humidity and air velocity in the chamber) and finally motion of the target screen. Morphological changes can occur upon decreasing the distance between the syringe needle and the substrate. Increasing the distance or decreasing the electrical field decreases the bead density, regardless of the concentration of the polymer in the solution. Elemental carbon in the sp^2 hybridization can form a variety of amazing structures. The nanotubes consisted of up to several tens of graphitic shells (so called multiwalled carbon nanotubes (MWNT)) with adjacent shell separation of 0.34 nm, diameters of 1 nm and high length/diameter ratio. Two years later, Iijima and Ichihashi synthesized SWNT. There are two main types of carbon nanotubes that can have high structural perfection. Single walled nanotubes (SWNT) consist of a single graphite sheet seamlessly wrapped into a cylindrical tube. MWNT comprise an array of such nanotubes that are concentrically nested like rings of a tree trunk [54–57].

Recent discoveries of fullerene, a zero dimensional form of carbon and carbon nanotube, which is a one-dimensional form, have stimulated great interest in carbon materials overall. Fullerenes are geometric cage-like structures of carbon atoms that are composed of hexagonal and pentagonal faces. When a Bucky ball is elongated to form a long and narrow tube of few nanometers diameter approximately which is the basic form of carbon nanotube. This stimulated a frenzy of activities in properties measurements of doped fullerenes. The discovery of fullerenes led to the discovery of carbon nanotubes by Iijima in 1991 [58–62]. The discovery of carbon nanotubes created much excitement and stimulated extensive research into the properties of nanometer scale cylindrical carbon networks. Many researchers have reported mechanical properties of carbon nanotubes that exceed those of any previously existing materials. Although there are varying reports in the literature on the exact properties of carbon nanotubes, theoretical and experimental results have shown extremely high modulus, greater than 1 TPa (the elastic modulus of diamond is 1.2 TPa) and reported strengths 10–100 times higher than the strongest steel at a fraction of the weight. Indeed, if the reported mechanical properties are accurate, carbon nanotubes may result in an entire new class of advanced materials.

In addition to the exceptional mechanical properties associated with carbon nanotubes, they also possess superior thermal and electric properties such as thermally stable up to 2800 °C in vacuum, thermal conductivity about twice as high as diamond, electric current carrying capacity 1000 times higher than copper wires. These exceptional properties of carbon nanotubes have been investigated for devices such as field emission displays, scanning probe microscopy tips and microelectronic devices. Carbon nanotubes present significant opportunities to basic science and nanotechnology, and pose significant challenge for future work in this field. The approach of direct growth of nano wires into ordered structures on surfaces is a promising route to approach nanoscale problem and create novel molecular scale devices with advanced electrical, electromechanical and chemical functions [54].

1.6.1   STRUCTURE AND PROPERTIES

1.6.1.1   ELECTRICAL PROPERTIES

The Unique Electrical Properties of carbon nanotubes are to a large extent derived from their 1-D character and the peculiar electronic structure of graphite. They have extremely low electrical resistance. Resistance occurs when an electron collides with some defect in the crystal structure of the material through which it is passing. The defect could be an impurity atom, a defect in the crystal structure, or an atom vibrating. Such collisions deflect the electron from its path, but the electrons inside a carbon nanotube are not so easily scattered. Because of their very small diameter and huge ratio of length to diameter a ratio that can be up in the millions or even higher. In a 3-D conductor, electrons have plenty of opportunity to scatter, since they can do so at any angle. Any scattering gives rise to electrical resistance. In a 1-D conductor, however, electrons can travel only forward or backward. Under these circumstances, only backscattering (the change in electron motion from forward to backward) can lead to electrical resistance. But backscattering requires very strong collisions and is thus less likely to happen. So the electrons have fewer possibilities to scatter. This reduced scattering gives carbon nanotubes their very low resistance. In addition, they can carry the highest 30 current density of any known material, measured as high as $109A/cm^2$. One use for nanotubes that has already been developed is as extremely fine electron guns, which could be used as miniature cathode ray tubes (CRTs) in thin high-brightness low-energy low-weight displays. This type of display would consist of a group of many tiny CRTs, each providing the electrons to hit the phosphor of one pixel, instead of having one giant CRT whose electrons are aimed using electric and magnetic fields. These displays are known as Field Emission Displays(FEDs). A nanotube formed by joining nanotubes of two different diameters end to end can act as a diode, suggesting the possibility of constructing electronic computer circuits entirely out of nanotubes. Nanotubes have been shown to be superconducting at low temperatures [55].

1.6.1.2   MECHANICAL PROPERTIES

The carbon nanotubes are expected to have high stiffness and axial strength as a result of the carbon–carbon sp² bonding. The practical application of the nanotubes requires the study of the elastic response, the inelastic behavior and buckling, yield strength and fracture. Efforts have been applied to the experimental and theoretical investigation of these properties. Nanotubes are the stiffest known fiber, with a measured Young's modulus of 1.4TPa. They have an expected elongation to failure of 20–30%, which combined with the stiffness, projects to a tensile strength well above 100Gpa (possibly higher), by far the highest known. For comparison, the Young's modulus of high-strength steel is around 200GPa, and its tensile strength is 1–2GPa [56].

1.6.1.3   THERMAL PROPERTIES

Prior to CNT, diamond was the best thermal conductor. CNT have now been shown to have a thermal conductivity at least twice that of diamond. CNT have the unique property of feeling cold to the touch, like metal, on the sides with the tube ends exposed, but similar to wood on the other sides. The specific heat and thermal conductivity of carbon nanotube systems are determined primarily by phonons. The measurements yield linear specific heat and thermal conductivity above 1K and below room temperature while a 0.62 behavior of the specific heat was observed below 1K. The linear temperature dependence can be explained with the linear k-vector dependence of the frequency of the longitudinal and twist acoustic phonons. The specific heat below 1^K can be attributed to the transverse acoustic phonons with quadratic k dependence. The measurements of the thermoelectric power (TEP) of nanotube systems give direct information for the type of carriers and conductivity mechanisms [57].

1.6.2   STRUCTURE OF CARBON NANO TUBES

The multilayered nanotubes were found in the cathode tip deposits that form when a DC arc is sustained between the graphite electrodes of a fullerene generator. They are typically composed of 2 to 5 concentric cylindrical shells, with outer diameter typically a few tens of nanometer and lengths of the order of micrometer. Each shell has the structure of a rolled up graphene sheet with the sp^2 carbons forming a hexagonal lattice. The discovery of nanotubes has revolutionized researches in different directions. A light and high strength nanotube would be an ideal structural member for designing nano structural instruments. It has reported that nanotubes could become as familiar as silicon in this century and the full development of the nanotubes would be around 2010. In order to familiarize the uses and applications of the nanotubes and their related products, an understanding of the structure, characterization and properties of the nanotubes is essential. The nanotubes possess conducting properties ranging from metallic to moderate band gap semiconductor. In general, the nanotubes could be specified in terms of the tube diameter (d) and the chiral angle (θ). The chiral vector (Ch) is defined as a line connected from two crystallographic ally equivalent sites on a two-dimensional graphene structure. The chiral vector can be defined in terms of the lattice translation indices (n, m) and the basic vectors a_1 and a_2 of the hexagonal lattice (a layer of grapheme sheet). The chiral angle (θ) is measured as an angle between the chiral vector Ch with respect to the zigzag direction (n, 0), where θ = 0°and the unit vectors of a_1 and a_2. The armchair nanotube is defined as the θ = 30° and the translation indices is (n, n). All other types of nanotubes could be identified as a pair of indices (n, m)where n ≠ m. The electronic conductivity is highly sensitive to a slight change of these parameters, which cause a changing of materials between metallic and semiconductor statues. Recently, it has been reported that the scanning tunneling microscopy (STM) and spectroscopy could be used to observe the electronic properties and atomic arrangement of SWNTs [55–57].

1.6.2.1   MULTI AND SINGLE WALL NANO TUBES

Multi-walled carbon nanotubes were first reported by Iijima in 1991 [55] in carbon made by an arc-discharge method4. About two years later, he made the observation of SWNTs. A SWNT is a grapheme sheet rolled over into a cylinder with typical diameter of the order of 1.4 nm, similar to that of a C_{60} buck ball. A MWNT consists of concentric cylinders with an interlayer spacing of 3.4Å and a diameter typically of the order of 10–20 nm. The lengths of the two types of tubes can be up to hundreds of microns or even centimeters. A SWNT is a molecular scale wire that has two key structural parameters. By folding a graphene sheet into a cylinder so that the beginning and end of a (m, n) lattice vector in the graphene plane join together, one obtains an (m, n) nanotube. The (m, n) indices determine the diameter of the nanotube, and also the so-called chirality.

1.6.3   SYNTHESIS METHODS OF CNT

1.6.3.1   SYNTHESIS OF CNT

The MWNT were first discovered in the soot of the arc-discharge method by Iijima. This method had been used long before in the production of carbon fibers and fullerenes. It took two more years for Iijima and Ichihashito synthesize SWNT by use of metal catalysts in the arc-discharge method in 1993. Significant progress was achieved by laser ablation synthesis of bundles of aligned SWNT with small diameter distribution. Catalytic growth of nanotubes by the chemical vapor decomposition (CVD) method was used too [54].

1.6.3.1.1   ARC-DISCHARGE

In 1991, Iijima reported the preparation of a new type of finite carbon structures consisting of needlelike tubes. The tubes were produced using an arc discharge evaporation method similar to that used for the fullerene synthesis. The carbon needles, ranging from 4 to 30 nm in diameter and up to 1 mm in length, were grown on the negative end of the carbon electrode used for the direct current (dc) arc-discharge evaporation of carbon in an argon filled vessel (100Torr). Iijima used an arc-discharge chamber filled with a gas mixture of 10Torr methane and 40Torr argon. Two vertical thin electrodes were installed in the center of the chamber. The lower electrode, the cathode, had a shallow dip to hold a small piece of iron during the evaporation. The arc-discharge was generated by running a dc current of 200A at 20 V. Laser beam vaporizes target of a mixture of graphite and metal catalyst (Co, Ni) in a horizontal tube in a flow of inert gas at controlled pressure and in a tube furnace at 1200 °C. The nanotubes are deposited on water-cooled collector outside the furnace electrodes. The use of the three components argon, iron and methane, was critical for the synthesis of SWNT. The nanotubes had diameters of 1 nm with a broad diameter distribution between 0.7 and 1.65 nm. In the arc-discharge synthesis of nanotubes, used as anodes thin electrodes with bored holes, which were filled with a mixture of pure powdered metals (Fe, Ni or Co) and graphite. The electrodes were vaporized with a current of 95–105A in 100–500Torr of He. Large quantities of SWNT were generated by the arc-technique. The arc was generated between two graphite electrodes in a reactor under helium atmosphere (660 mbar)[56].

1.6.3.1.2   LASER-ABLATION

In 1996, Smalley et al. [56] produced high yields (>70%) of SWNT by laser ablation (vaporization) of graphite rods with small amounts of Ni and Co at 1200 °C. The tube grows until too many catalyst atoms aggregate on the end of the nanotube. The large particles either detach or become over coated with sufficient carbon to poison the catalysis. This allows the tube to terminate with a fullerene like tip or with a catalyst particle. Both arc-discharge and laser-ablation techniques have the advantage of high (>70%) yields of SWNT and the drawback that (1) they rely on evaporation of carbon atoms from solid targets at temperatures >3000 °C, and(2) the nanotubes are tangled which makes difficult the purification and application of the samples.

1.6.3.1.3   CHEMICAL VAPORS DEPOSITION (CVD)

Despite the described progress of synthetic techniques for nanotubes, there still remained two major problems in their synthesis, that is, large scale production and ordered synthesis. But, in 1996a CVD method emerged as a new candidate for nanotube synthesis. This method is capable of controlling growth direction on a substrate and synthesizing 13a large quantity of nanotubes. In this process a mixture of hydrocarbon gas, acetylene, methane or ethylene and nitrogen was introduced into the reaction chamber. During the reaction, nanotubes were formed on the substrate by the decomposition of the hydrocarbon at temperatures 700–900 °C and atmospheric pressure. The process has two main advantages: the nanotubes are obtained at much lower CVD reactor temperature, although this is at the cost of lower quality, and the catalyst can be deposited on a substrate, which allows for the formation of novel structures [56].

1.6.3.1.4   THE SUBSTRATE

The preparation of the substrate and the use of the catalyst deserve special attention, because they determine the structure of the tubes. The substrate is usually silicon, but also, glass and alumina are used. The catalysts are metal nanoparticles, like Fe, Co and Ni, which can be deposited on silicon substrates either from solution, electron beam evaporation or by physical sputtering. The nanotube diameter depends on the catalyst particle size, therefore, the catalyst deposition technique, in particular the ability to control the particle size, is critical to develop nano devices. Porous silicon is an ideal substrate for growing self-oriented nanotubes on large surfaces. It has been proven that nanotubes grow at a higher ratio (length per minute), and they are better aligned than on plain silicon. The nanotubes grow parallel to each other and perpendicular to the substrate surface, because of catalyst surface interaction and the Vander Waals forces developed between the tubes [56].

1.6.3.1.5   THE SOL-GEL

The sol-gel method uses a dried silicone gel, which has undergone several chemical processes, to grow highly aligned nanotubes. The substrate can be reused after depositing new catalyst particles on the surface. The length of the nanotube arrays increases with the growth time, and reaches about 2 mm after 48h growth [56].

1.6.3.1.6   GAS PHASE METAL CATALYST

In the methods described above, the metal catalysts are deposited or embedded on the substrate before the deposition of the carbon begins. A new method is to use a gas phase for introducing the catalyst, in whichboth the catalyst and the hydrocarbon gas are fed into a furnace, followed by catalytic reaction in the gas phase. The latter method is suitable for large-scale synthesis, because the nanotubes are free from catalytic supports and the reaction can be operated continuously. A high-pressure carbon monoxide (CO) reaction method, in which CO gas reacts with iron penta carbonyl, $Fe(CO)_5$ to form SWNT, has been developed. SWNT have also been synthesized from a mixture of benzene and ferrocene, $Fe(C_5H_5)_2$ in a hydrogen gas flow. In both methods, catalyst nanoparticles are formed through thermal decomposition of oregano metallic compounds, such as iron penta carbonyl and ferrocene. The reverse micelle method is promising, which contains catalyst nanoparticles (Mo and Co) with a relatively homogeneous size distribution in a solution. The presence of surfactant makes the nanoparticles soluble in an organic solvent, such as toluene and benzene. The colloidal solution can be sprayed into a furnace, at a temperature of 1200 °C; it vaporizes simultaneously with the injection and a reaction occurs to form a carbon product. The toluene vapor and metal nanoparticles act as carbon source and catalyst, respectively. The carbon product is removed from the hot zone of the furnace by a gas stream (hydrogen) and collected at the bottom of the chamber [55, 56].

1.6.3.2 RECENT TRENDS IN THE SYNTHESIS OF CNT

Some researchers synthesized carbon nanotubes from an aerosol precursor. Solutions of transition metal cluster compounds were atomized by electro hydrodynamic means and the resultant aerosol was reacted with ethane in the gas phase to catalyze the formation of carbon nanotubes. The use of an aerosol of iron penta carbonyl resulted in the formation of multiwalled nanotubes, mostly 6–9 nm in diameter, whereas the use of iron do deca carbonyl gave results that were concentration dependent. High concentrations resulted in a wide diameter range (30–200 nm)whereas lower concentrations gave multiwalled nanotubes with diameters of 19–23 nm. CNT synthesized by electrically arcing carbon rods in helium (99.99%) in a stainless steel chamber with an inner diameter of 600 mm and a height of 350 mm. The anode was a coal-derived carbon rod (10 mm in diameter, 100–200 mm in length); the cathode was a high-purity graphite electrode (16 mm in diameter, 30 mm in length). The helium gas functioned as buffer gas and its pressure was varied in range of 0.033–0.076 MPa in the experiment. CNT synthesized via a novel route using an iron catalyst at the extremely low temperature of 180 °C. The carbon clusters can grow into nanotubes in the presence of Fe catalyst, which was obtained by the decomposition of iron carbonyl $Fe_2(CO)_9$ at 250 °C under nitrogen atmosphere. SWNT have been successfully synthesized using a fluidized bed method that involves fluidization of a catalyst/support at high temperatures by a hydrocarbon flow. A new method, which combines none equilibrium plasma reaction with template controlled growth technology, has been developed for synthesizing aligned carbon nanotubes at atmospheric pressure and low temperature. Multiwall carbon nanotubes with diameters of approximately 40 nm were restrictedly synthesized in the channels of anodic aluminum oxide template from a methane hydrogen mixture gas by discharge plasma reaction at a temperature below 200 °C. In a recent technique, Nebulized spray pyrolysis, large-scale synthesis of MWNT and aligned MWNT bundles is reported. Nebulized spray is a spray generated by an ultrasonic atomizer. A SEM images of aligned MWNT bundles obtained by the pyrolysis of a nebulized spray of ferrocene–toluene–acetylene mixture. The advantage of using a nebulized spray is the ease of scaling into an industrial scale process, as the reactants are fed into the furnace continuously [55, 56].

During nanotube synthesis, impurities in the form of catalyst particles, amorphous carbon and nontubular fullerenes are also produced. Thus, subsequent purification steps are needed to separate the carbon nanotubes. The gas phase processes tend to produce nanotubes with fewer impurities and are more amenable to large scale processing. It is believed that the gas phase techniques, such as CVD, for nanotube growth offer greater potential for the scaling up of nanotubes production for applications. Initially, electric-arc discharge technique was the most popular technique to prepare the SWNTs as well as MWNTs. In this technique, the carbon arc provides a simple and traditional tool for generating the high temperatures needed for the vaporization of carbon atoms into a plasma (>3000 °C).The gas phase growth of single walled nanotubes by using carbon monoxide as the carbon source has already been reported. They found the highest yields of single walled nanotubes occurred at the highest accessible temperature and pressure (1200 °C, 10atm). They have modified this process to produce large quantities of single walled nanotubes with remarkable purity. The lower processing temperatures also enable the growth of carbon nanotubes on a wide variety of substrates. CVD method has been successful to produce the CNTs in large quantity, and also to obtain the vertically aligned CNTs at relatively low temperature. In particular, growth of vertically aligned CNTs on large substrate area at low temperature, for instance, softening temperature of the glass is an important factor for the practical application of the electron emitters to the field emission displays. A lot of reports on the synthesis of single-walled as well asmultiwalled carbon nanotubes using the plasma enhanced chemical vapor deposition and microwave plasma enhanced chemical vapor deposition techniques, are available in the literature. Others successfully synthesized vertically aligned CNTs at 550 °C on Ni-coated Si substrate placed parallel to Pd plate as a dual catalyst and tungsten wire filament. The bamboo shaped carbon nanotubes can be obtained only if the reaction temperature is higher than 1000K, and carbon fibers can be obtained at lower temperatures. They have also discussed the role and state of the catalyst particles. They have found that a plug-shaped Ni particle always plunged at the top end of a nanotube. Experimental results indicated that the catalytic particles exist in a liquid state during the synthesis procedure. After having crystallized, the orientations of plunged Ni particles randomly distributed around the axis. All three metals deposited on the quartz plates are found to be efficient catalysts for the growth of CNTs in good alignment by CVD using ethylene diamine [56, 57].

The CNTs are multiwalled with a bamboo-like graphitic structure. The hollow compartments of the tubes are separated by the graphitic interlink layers. This is a simple and efficient way for the production of carbon nanotubes with good order using ethylene diamine by CVD without plasma aid. Others produced CNTs by hotwire CVD using a mixture of C_2H_2 and NH_3 gases. They used crystalline Si and SiO_2 substrates coated with Ni films of 30 nm in thickness as a catalyst for nanotube formation. SEM images showed that micron sized grains were present in the deposit on the Si substrate, whereas nano sized grains were evident in the deposit on the SiO_2 substrate. Nanotube formation could not be confirmed by SEM, but there was evidence of the possible formation of nanotubes on the Ni-coated SiO_2 substrate. Two novel nanostructures, which are probably the missing link between onion like carbon particles and nanotubes, have also been obtained. Low synthesis temperature <520°Cdue to the nonequilibrium characteristics of microwave plasma operated at low pressure is also reported, which is crucial for some fascinating applications. Some researchers employed a high-density plasma chemical vapor deposition (PECVD) to grow high quality carbon nanotubes at low temperatures. High density, aligned CNTs can be grown on Si and glass substrate. The

CNTs were selectively deposited on the patterned Ni catalyst layer, which was sputtered on Si substrate. Some others synthesized pure carbon nanotubes at very low temperature using MW-PECVD with methane/hydrogen gas. Others prepared the massive carbon nanotubes on silicon, quartz and ceramic substrates using MW-PECVD. The nanotubes, ranging from 10 to 120 nm in diameter and a few tens of microns in length, were formed under hydrocarbon plasma at 720 °C with the aid of iron-oxide particles. Morphology of the nanotubes is strongly influenced by the flow ratio of methane to hydrogen. Defect less nanotubes with small diameters are favorably produced under a small flow ratio. To date, many methods for synthesizing carbon nanotubes have been developed and most of which operated at high temperature over 4000 °C for graphite arc discharge and laser ablation. These temperatures are too high and unsuitable for the fabrication of electronic devices because most electric connections are made of aluminum with the melting point below 700 °C. These challenges have promoted the current exploration of low temperature synthesis of carbon nanotubes such as CVD, PECVD and MWPECVD. Recently, these low temperature methods have been successful in growing highly aligned and large quantities of carbon nanotubes. It is also possible to control over length, diameter and structure of carbon nanotubes grown by CVD techniques. Therefore, CVD techniques are the most popular methods to synthesize the carbon nanotubes [56–182].

Recently have grown straight carbon nanotubes, carbon nanotubes "knees," Y-branches of carbon nanotubes and coiled carbon nanotubes on a graphite substrate held at room temperature by the decomposition of fullerene under moderate heating at 450 °C in the presence of 200 nm Ni particles. The grown structures were investigated without any further manipulation by STM. The formation of the carbon nanostructures containing nonhexagonal rings is attributed partly to the tem plating effect of the high pyrolitic graphite (HOPG), partly to the growth at room temperature, which enhances the probability of quenching-in for nonhexagonal rings. Similar coiled nanotubes were found after several steps of chemical treatment in a catalytically grown carbon nanotube sample. They have examined growth of CNTs on a porous alumina template in order to improve the selectivity and uniformity of CNTs. They have fabricated well ordered, nano sized pores on a Si substrate using an anodic oxidation method. Recently developed a simple process for selective removal of carbon from single walled carbons nanotube samples based on a mild oxidation by carbon dioxide. Some nanotubes were found to be partially filled with a solid material (probably metallic iron) that seems to catalyze the nanotube growth. Some regions of the deposit also revealed the presence of nanoparticles. The present experimental conditions should be suitable to produce locally structured deposits of carbon nanotubes for various applications. Recently synthesized straight and bamboo like carbons nanotubes in a methane diffusions flame using a Ni–Cr–Fe wire as a substrate. The catalyst particles were nickel and iron oxides formed on the wire surface inside the flame. The carbon growth over the catalysts has been followed gravimetrically in situ. The reaction was stopped after different pulse numbers in an attempt to control both the diameters and the lengths of the carbon nanotubes, which were characterized by transmission electron microscopy (TEM). Recently prepared two kinds of catalytic layers onto n-typed silicon substrate nickel by sputtering and iron(III) nitrate metal oxides by spin coatings. For iron(III) nitrate metal oxide 0.5 mol of ferric nitrate non hydrate (Fe$_2$(NO$_3$)$_2$.9H$_2$O] ethanol solution was coated onto tubes on both Ni and iron(III) nitrate metal oxide layers by the HFPECVD (hot filament plasma enhanced chemical vapor deposition) method [55–58].

1.6.4 GRAFTING OF POLYMERS

The covalent reaction of CNT with polymers is important because the long polymer chains help to dissolve the tubes into a wide range of solvents even at a low degree of fictionalization. There are two main methodologies for the covalent attachment of polymeric substances to the surface of nanotubes, which are defined as "grafting to" and "grafting from" methods. The former relies on the synthesis of a polymer with a specific molecular weight followed by end group transformation. Subsequently, this polymer chain is attached to the graphitic surface of CNT. The "grafting from "method is based on the covalent immobilization of the polymer precursors on the surface of the nanotubes and subsequent propagation of the polymerization in the presence of monomer species [182].

1.6.4.1 "GRAFTING TO" METHOD

Recently reported the chemical reactions of CNT and PMMA using ultrasonications. The polymer attachment was monitored by FT-IR and TEM.

As a result of this grafting, CNT were purified by filtration from carbonaceous impurities and metal particles. A nucleophilic reaction of polymeric an alternative approach was reported by the group of researchers MWNT were functionalized with n-butyl lithium and subsequently coupled with halogenated polymers. Microscopy images showed polymer-coated tubes while the blend of the modified material and the polymer matrix exhibited enhanced properties in tensile testing experiments. They reported the grafting of functionalized polystyrene to CNT via a cyclo addition reaction. An azido polystyrene with a defined molecular weight was synthesized by atom transfer radical polymerization and then added to nanotubes. In a different approach, chemically modified CNT with appended double bonds were functionalized with living poly styryl lithium anions via anionicpolymerization.124b. The resulting composites were soluble in common organic solvents. Using an alternative method, polymers prepared by nit oxide-mediated free radical polymerization were used to functionalize SWNT through a radical coupling reaction of polymer-centered radicals. The in situ generation of polymer radical species takes place via thermal loss of the nitroxide-capping agent. The polymer-grafted tubes were fully characterized by UV-vis, NMR, and Raman spectroscopy [182].

1.6.4.2   "GRAFTING FROM" METHOD

CNT-polymer composites were first fabricated by an in situ radical polymerization process. Following this procedure, the double bonds of the nanotube surface were opened by initiator molecules and the CNT surface played the role of grafting agent. Similar results were obtained by several research groups. Depending on the type of monomer, it was possible not only to solubilize CNT but also to purify the raw material from catalyst or amorphous carbon. Through the negative charges of the polymer chain, the composite could be dispersed in aqueous media, whereas the impurities were eliminated by centrifugation. In a subsequent work, the same authors fabricated films consisting of alternating layers of anionic PSS-grafted nanotubes and cationic diazo polymer. The ionic bonds in the film were converted to covalent bonds upon UV irradiation, which improved greatly the stability of the composite material. They prepared poly vinyl pyridine (PVP)-grafted SWNT by in situ polymerization. Solutions of such composites remained stable for at least 8 months. Layer by layer deposition of alternating thin films of SWNT-PVP and poly (acrylic acid) resulted in freestanding membranes, held together strongly by hydrogen bonding. Assemblies of PSS-grafted CNT with positively charged porphyries were prepared via electrostatic interactions. Then an assembly gave rise to photo induced intra complex charge separation that lives for tens of microseconds. The authors have demonstrated that the incorporation of CNT-porphyrin hybrids onto indium tin oxide (ITO) electrodes leads to solar energy conversion devices. The raw material was treated with sec-butyl lithium, which introduces a carbon ionic species on the graphitic surface and causes exfoliation of the bundles. When a monomer was added, the nanotube carbon ions initiate polymerization, resulting in covalent grafting of the polystyrene chains. They studied the fabrication of composites by in situ ultrasonic induced emulsion polymerization of acryl ate. It was not necessary to use any initiating species, and the polymer chains were covalently attached to the nanotube surface. MWNT grafted with poly (methyl methacrylate) were synthesized by emulsion polymerization of the monomer in the presence of a radical initiator 130b or a cross-linking agent. CNT were found to react mostly with radical-type oligomers. The modified tubes had an enhanced adhesion to the polymer matrix, as could be observed by the improved mechanical properties of the composite. A different approach to composite preparation involves the attachment of atom transfer radical polymerization (ATRP) initiators to the graphitic network. These initiators were found to be active in the polymerization of various acryl ate monomers. Recently prepared and characterized composites of nanotubes with methyl methacrylate and tetra-butyl acryl ate. The former composites were found to be insoluble in common solvents, while the latter were soluble in a variety of organic media. The fabrication of nanotube poly aniline composites via in situ chemical polymerization of aniline was studied by many groups. Initially, a charge transfer interaction was suggested, whereas a covalent attachment between the two components was described. The surface modification of SWNT was reported recently via in situ Ziegler-Natta polymerization of ethylene. The exact mechanism of nanotube-polymer interaction remains unclear, althoughthe authors suggested that a possible cross-linking could take place between the two components. The development of an integrated nanotube-epoxy polymer composite was reported by some authors. In the fabrication process, the authors used functionalized tubes with amino groups at the ends. These moieties could react easily with the epoxy groups and act as curing agents for the epoxy matrix. The cross-linked structure was most likely formed through covalent bonds between the tubes and the epoxy polymer. Multi-walled CNT were successfully modified with polyacrylonitrile chains by applying electrochemical polymerization of the monomer. The surface-functionalized tubes showed a good degree of dispersion in DMF while further proofs of de bundling were obtained by TEM images [56, 182].

1.7   CNT-COMPOSITES BASED ADSORBENTS

Potential practical applications of CNTs such as chemical sensors, field emission, electronic devices, high sensitivity nano balance for nanoscopic particles, nano tweezers, reinforcements in high performance composites, biomedical and chemical investigations, anode for lithium ion batteries, super capacitors and hydrogen storage have been reported. Even though the challenges in fabrication may prohibit realization of many of these practical device applications, the fact that the properties of CNTs can be altered by suitable surface modifications can be exploited for more imminent realization of practical devices. In this respect, a combination of CNTs and other nanomaterials, such as nano crystalline metal oxide/CNTs, polymer/CNTs and metal filled CNTs may have unique properties and research have therefore been focused on the processing of these CNT based nano composites and their different applications [54–57].

Adsorption of single metal ions, dyes and organic pollutants on CNT-based adsorbent composites is one of the most applications of these materials. MWCNT/iron oxide magnetic composites were prepared and used for adsorptions of several metal ions. The CNTs were purified by using nitric acid, which results in modification of the surface of the nanotubes with oxygen containing groups like carbonyl and hydroxyl groups. The adsorption capability of the composite is higher than that of nanotubes and activated carbon. The sorption of ions such as Pb(II) and Cu(II) ions on the composite were spontaneous and endothermic processes based on the thermodynamic parameters (ΔH, ΔS and ΔG) calculated from temperature dependent sorption isotherms. Alumina-coated MWCNTs were synthesized and reported for its utilization as adsorbent for the removal of lead ions from aqueous solutions in two modes. With an increase in influent pH between 3 and 7, the percentage of lead removed increases. The adsorption capacity increases by increasing agitation speed, contact time and adsorbent dosage. The reported composite can be regenerated as it was confirmed by SEM and EDX analysis [58–60].

Grafting of polymers to nanotubes has been realized via both"grafting-to" and "grafting-from" approaches that mention before. The grafting-to method is based on attachment of premade end-functionalized polymer onto the tips and convex walls of the nanotubes via chemical reactions such as etherification and amidization. One advantage of this method is that the mass and distribution of the grafted polymers can be more precisely controlled. However, initial grafted polymer chains sterically prevent diffusion of additional polymer chains to the nanotube surface resulting in low grafting density. The grafting-from method is based on immobilization of reactive groups (initiators) onto the surface, followed by in situ polymerization of appropriate monomers to form polymer-grafted nanotubes. The advantage of this approach is that very high grafting density can be achieved. But careful control of the amount of initiator and the conditions for the polymerization reaction is required. Rather than the grafting-to method used by some researchers. They investigated the grafting-from approach, which involves the propagation of dendrimers from CNT surfaces by in situ polymerization of monomers in the presence of CNT attached macro initiators [73].

The mobility and small size of the fugitive monomers, in contrast to that of preformed dendritic polymers, are expected to improve the efficiency and yield of the grafting-from approach and the resulting de bundling process. The number of peripheral reactive groups can be precisely controlled by choosing the appropriate synthetic generation. Fictionalization of nanotubes with dendrimers represents a particularly promising strategy to attach an in principle unlimited number of functional groups onto the SWNT surfaces, and thus to significantly increase the compatibility and reactivity of CNTs with thermosetting polymer matrices, such as epoxy, bismaleimide, and cyan ate ester. Finally, to achieve good alignment of nanotubes to exploit their superior anisotropic mechanical properties, they applied the reactive spinning process. Spinning has been widely used for thermoplastic resins such as polyvinyl acetate (PVA), polyamide (PA), and polycarbonate (PC), but it has rarely been applied to thermosetting resins, except for cyan ate ester. Control of viscosity within a narrow range suitable for spinning is often tricky and resin-specific. By comparison with cyan ate ester, thermosetting epoxy generally has a smaller range of spinning conditions due to its higher reactivity. However, use of thermo set resins as the matrix for fiber spinning offers advantages of lower viscosity and reaction with functional groups on the nanotubes [72, 182].

Although, several reports demonstrated that CNTs have good adsorption capacities for dyes due to their hollow and layered nano sized structures, which in turn have a large specific surface area, but their high cost restricts their use from industrial application. Furthermore, separation of CNTs from aqueous solution is very difficult because of its smaller size and high aggregation property. The problem of cost and separation of CNTs from aqueous medium can be overcome by making composites of CNTs with polymers, metal oxide, carbon etc. Such materials act as a stable matrix to the CNTs. Recently the CNTs have been used as a promising nano filler for the preparation of CNT based nano composites because of their excellent improved adsorption, mechanical, electrical and thermal properties. CNT based composites are expected to be excellent adsorbent because CNTs provide not only the additional active sites but also larger surface area, which in turn makes them good adsorbents compared to their parent materials [72] (Figs. 1.1 and 1.2).

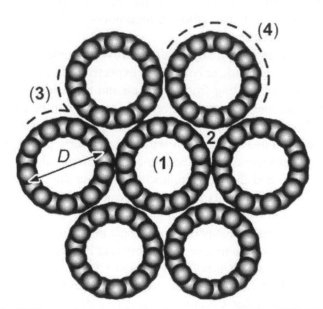

FIGURE 1.1   Different adsorption sites in a homogeneous bundle of SWCNTs with tube diameter D: (1) intra tubular, (2) interstitial channel, (3) external groove, and(4) exposed surface of peripheral tube. Sites 1 and 2 comprise the internal porous volume of the bundle, whereas sites 3 and 4 are both located on the external surface of the bundle.

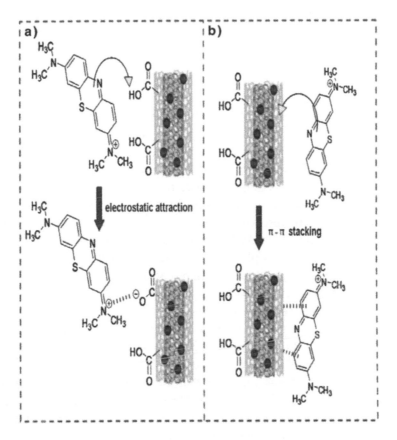

FIGURE 1.2 Schematic illustration of the possible interaction between MWCNTs and methylene blue: (a) electrostatic attraction and (b) π–π stacking

CNTs can be combined with various metal oxides for the degradation of some organic pollutants too. Carbon nanotubes/metal oxide (CNT/MO) composites can be prepared by various methods such as wet chemical, sol gel, physical and mechanical methods. To form nanocomposite, CNTs can be combined with various metal oxides like Ti_2O_3, ZnO, WO_3, Fe_2O_3, and Al_2O_3. The produced nano composite can be used for the removal of various pollutants. Nanoscale Pd/Fe particles were combined with MWNTs and the resulted composite was used to remove 2, 4-dichlorophenol (2, 4-DCP). It was reported that the MB adsorption was pH-dependent and adsorption kinetics was best described by the pseudo-second-order model.

Iron oxide/CNT composite was reported to be efficient adsorbent for remediation of chlorinated hydrocarbons. The efficiency of some other nano composites like CNT/alumina, CNT/titanium and CNT/ZnO has also been reported [60–62].

Chitosan (CS) is one of the best adsorbents for the removal of dyes due to its multiple functional groups, biocompatibility and biodegradability, but its low mechanical strength limits its commercial applications. Impregnation of CS hydro gel beads with CNTs (CS/CNT beads) resulted in significant improved mechanical strength. In CS/CNT composite, CNTs and CS are like a symbiosis, CNTs help to improve the mechanical strength of CS while CS help to reduce the cost of CNTs for adsorption, while the resulted composite solves the problem of separating CNTs from aqueous medium. To resolve the aggregation and dispersion problem of CNTs, prepared the CNTs/ACF composite and its application was investigated for the removal of phenol and basic violet 10(BV10). CNTs/ACF was prepared via directly growing nano scaled CNTs on micro scaled carbon matrix. Poly acrylonitrile was used as a source of carbon. From the results, it was observed that dye adsorption equilibrium time for CNTs/ACF is shorter as compared to ACF and monolayer adsorption capacity does not display a linear increase with increasing the BET surface area. The decoration of CNTs tends to lower the porosity of the ACF from 1065 to 565 m^2/g. This finding indicates that the total micro porosity of ACF cannot be fully accessed by the dye molecules. Therefore, the appearance of CNTs plays a positive role in (i) facilitating the pore accessibility to adsorbents and (ii) providing more adsorptive sites for the liquid-phase adsorption. This reflects that CNTs/ACF contains a large number of mesopore channels, thus preventing the pore blockage from the diffusion path of microspores for adsorbents to penetrate [72, 73].

Incorporation of magnetic property in CNTs is another good technique to separate CNTs from solution. The magnetic adsorbent can be well dispersed in the water and easily separated magnetically Magnetic-modified MWCNTs were used for removal of cationic dye crystal violet (CV), thionine (Th), Janus green B (JG), and Methylene Blue (MB).To find the optimum adsorption, effect of various parameters including initial pH, dosages of adsorbent and contact time have been investigated. The optimum adsorption was found to be at pH 7.0 for all dyes. The removal efficiency of cationic dyes using GG/MWCNT/Fe_3O_4 is higher as compared with other adsorbents such as MWCNTs and MWCNT/Fe_3O_4. The magnetic

GG/MWCNT/Fe$_3$O$_4$possesses the high adsorption properties and magnetic separation and can therefore be used as magnetic adsorbents to remove the contaminants from aqueous solutions. A novel magnetic composite bio-adsorbent composed of chitos and wrapping magnetic nano sized γ-Fe$_2$O$_3$ and multiwalled carbon nanotubes (m-CS/γ-Fe$_2$O$_3$/MWCNTs) was prepared for the removal of methyl orange. The adsorption capacity of MO onto m-CS/γ-Fe$_2$O$_3$/MWCNTs was 2.2 times higher than m-CS/γ-Fe$_2$O$_3$. The adsorption capacity of MO onto m-CS/γ-Fe$_2$O$_3$/MWCNTs was also higher than MWCNTs. Kinetics data and adsorption isotherm data were better fitted by pseudo-second-order kinetic mode land by Langmuir isotherm, respectively. [58–62, 72, 73].

Recently developed novels adsorbents by inserting MWCNTs into the cavities of dolomites for scavenging of ethidium the foam lines CNTs/dolomite adsorbent. Foam like ternary composite PUF/diatomite/dispersed-MWCNTs, gave the highest capacities for adsorption of these dyes, followed by PUF/agglomerated-MWCNTs, and then PUF/dispersed-MW-CNTs. Adsorption isotherm study revealed the monolayer adsorption at higher concentration and multilayer adsorption at lower concentration. Pseudo first order kinetics gives the best-fitted results compared to the pseudo second order. Self-assembled cylindrical graphene–MCNT (G–MCNT) hybrid, synthesized by the one pot hydrothermal process was used as adsorbent for the removal of methylene blue in batch process. G–MCNT hybrid showed good performance for the removal [72, 73].

Most important CNT-based as adsorbent composites include: CNT–Chitosan, CNT-ACF, CNT-Dolomite, CNT–Cellulose, CNT–Magnetic and Metal Oxide composites, CNT–Fiber composites, CNT–CF(PAN), CNT–Alginate, CNT–PANI composites and CNTs–Graphene. These composites are investigated in next section.

1.7.1   CNT-CHITOSAN COMPOSITES

Chitosanis one of the best adsorbents for the removal of dyes due to its multiple functional groups, biocompatibility and biodegradability, but its low mechanical strength limits its commercial applications. Impregnation of CS hydro gel beads with CNTs (CS/CNT beads) resulted in significant improved mechanical strength. In CS/CNT composite, CNTs and CS are like a symbiosis, CNTs help to improve the mechanical strength of CS while CS help to reduce the cost of CNTs for adsorption, while the resulted composite solves the problem of separating CNTs from aqueous medium. Impregnation of0.05wt% cetyl tri methyl ammonium bromide (CTAB) increased the maximum adsorption capacity of CS beads from media. The small difference in maximum adsorption capacity of CS/CNT beads and CS/CTAB beads indicated that CTAB molecules played a significant role in enhancing the adsorption performance of both varieties of beads. However, higher maximum adsorption capacity of CS/CNT beads than CS/CTAB beads suggested that CNTs itself in the beads adsorb CR during adsorption. This chapter suggests that surfactant played an important role in the removal of dyes. In his further studies on CS/CNTs, some researchers reported the effect of anionic and cationic surfactants as dispersant on impregnated MWCNTs/CS for the removal of CR dye [72, 73].

The adsorption capacity of CNT-impregnated CSBs was found to be dependent on the nature of dispersant for CNTs. CNT-impregnated CSBs were prepared by four different strategies for dispersing CNTs: (a) in CS solution (CSBN1), (b) in sodium do decyl sulfate (SDS) solution (CSBN2), (c) in CS solution containing cetyl tri methyl ammonium bromide (CTAB) (CSBN3), and(d) in SDS solution for gelatins with CTAB-containing CS solution (CSBN4). The adsorption capacities of composite for CR were in the order of: CSBN4 > CSBN1 > CSBN2 > CSBN3. The adsorption capacity of CSBN1 was slightly higher than CSBN2 for CR. This could be due to the dispersion in SDS molecules resulting in negatively charged CNTs as a result of adsorption of SDS molecules onto CNTs. The CR adsorption onto CSBN4 increased as the concentration of CNTs increased because of better dispersion of CNTs in SDS solution than in CTAB solution. CSBN3 showed poor adsorption capacity because CNTs were dispersed in CTAB and aggregates of CNTs blocked the adsorption sites on CS, CTAB, and SDS molecules in the beads [72].

However, covalent functionalization of CNTs with polymers has proved to be an effective way to improve their dispersion stability and make the resulting composites more stable and controllable CS, a natural polysaccharide with similar structural characteristics to cellulose, obtained by the de acetylation of chitin. It is a biocompatible, biodegradable, andnontoxic natural biopolymer and has excellent film-forming abilities. CS can selectively adsorb some metal ions, and has been successfully used in wastewater treatment. However, because CS is very sensitive to the pH of ionic solutions, its applications are limited. Several strategies have been devised to prepare CS derivatives that are insoluble in acid solutions and to preserve the adsorption capacity of CS. In order to increase the chemical stability of CS in acid solutions and improve its metal-ion-adsorbing properties, Schiff base-chitosan (S-CS) was produced by grafting aldehydes onto the CS backbone. Although the preparation of CS-modified MW-CNTs via covalent interactions have been reported by a few groups, so far no work has been published on the application of MWCNTs covalently modified with CS derivatives to SPE column preconcentration for ICP-MS determination of trace metals. A novel material was synthesized by covalently grafting S-CS on to the surfaces of MWCNTs, and used for preconcentration of V (V), Cr(VI), Cu(II), As (V), andPb(II) in various samples, namely herring, spinach, river water, and tap water, using an SPE method. The results demonstrated that the proposed multi element enrichment method can be successfully used for analysis of V(V), Cr(VI), Cu(II), As (V), andPb(II) in environmental water and biological samples. The method is fast and has good sensitivity and excellent precision. Compared with previously reported procedures, the present method has high enrichment factors and sensitivity. In short, the proposed method is suitable for preconcentration and separation of trace/ultra-trace metal ions in real samples [189].

1.7.2 CNT-ACF AS ADSORBENT

To resolve the aggregation and dispersion problem of CNTs, prepared the CNTs/ACF)composite and its application was investigated for the removal of phenol and BV10. CNTs/ACF was prepared via directly growing nano scaled CNTs on micro scaled carbon matrix. Poly acrylonitrile was used as a source of carbon. From the results, it was observed that dye adsorption equilibrium time for CNTs/ACF is shorter as compared to ACF and monolayer adsorption capacity does not display a linear increase with increasing the BET surface area. The decoration of CNTs tends to lower the porosity of the ACF. This finding indicates that the total micro porosity of ACF cannot be fully accessed by the dye molecules. However, as-produced CNTs offer more attractive sites, including grooves between adjacent tubes on the perimeter of the bundles, accessible interstitial channels, and external nanotube walls. Therefore, the appearance of CNTs plays a positive role in (i) facilitating the pore accessibility to adsorbents and (ii) providing more adsorptive sites for the liquid-phase adsorption. This reflects that CNTs/ACF contains a large number of mesopore channels, thus preventing the pore blockage from the diffusion path of microspores for adsorbates to penetrate [72].

A novel technique was performed to grow high-density CNTs that attach to PAN-based ACFs. Such unique nano/micro scaled carbon composites can serve as an excellent electrode material of EDLCs. The existence of CNTs is believed to play two important roles in enhancing the performance of EDLCs: (i)since its has good electric conductivity, the presence of CNTs would promote the contact electron transfer or lower contact resistance between the current collector and the carbon composite;(ii) CNTs not only provide additional exterior surface area for double-layer formation but also shift from microspore size distribution to mesopore size distribution that may reduce ionic transfer resistance and improve the high rate discharge capability. It is generally recognized that transition metals (Fe, Co, and Ni) can serve as catalytic sites in inducing carbon deposition, thus forming CNTs. This reveals that the uniform dispersion of the "seeds" on ACF surface, followed by a catalytic chemical vapor deposition (CCVD) treatment, would offer a possibility to fabricate the unique nano/micro scaled carbon composites. The mesopores channels, which came from CNT branches, would provide available porosity that is accessible for ionic transport and energy storage. This chapter intends to investigate the applicability of using the carbon composite as an electrode material for EDLCs. Two configurations (with and without CNTs) have been compared with respect to their double-layer capacitances and high-rate capability, analyzed by CV and charge-discharge cycling. This chapter has demonstrated that the specific capacitance and high-rate capability of PAN-based ACF in H_2SO_4 can be enhanced with the decoration of CNTs. A CVD technique enabled to catalytically grow CNTs onto ACF, thus forming CNT-ACF composite. N_2 physisorption indicated that the mesopore fraction of ACF is found to increase after the growth of nanotubes. The specific double-layer capacitance and high-rate capability were significantly enhanced because of the presence of nanotubes. The distributed capacitance effect and inner resistance were significantly improved for the CNT-ACF capacitor. Owing to the decoration of CNTs, the specific capacitance was found to have an increase of up to 42% [190].

1.7.3 CNT-METAL OXIDES AND MAGNETIC COMPOSITES

Incorporation of magnetic property in CNTs is another good technique to separate CNTs from solution. The magnetic adsorbent can be well dispersed in the water and easily separated magnetically. Magnetic-modified MWCNTs were used for removal of cationic dye CV, Th, JG, and MB. To find the optimum adsorption, effect of various parameters including initial pH, dosages of adsorbent and contact time have been investigated. The optimum adsorption was found to be at pH 7.0 for all dyes. In recent work, they prepared guar gum grafted Fe_3O_4/MWCNTs (GG/MWCNT/Fe_3O_4) ternary composite for the removal of natural red (NR) and MB. The removal efficiency of cationic dyes using GG/MWCNT/Fe_3O_4 is higher as compared with other adsorbents such as MWCNTs and MWCNT/Fe_3O_4. The higher adsorption capacity of GG/MWCNT/Fe_3O_4 could be related to the hydrophilic property of GG, which improved the dispersion of GG–MWCNT–Fe_3O_4 in the solution, which facilitated the diffusion of dye molecules to the surface of CNTs. The magnetic GG/MWCNT/Fe_3O_4 possesses the high adsorption properties and magnetic separation and can therefore be used as magnetic adsorbents to remove the contaminants from aqueous solutions. Starch functionalized MWCTs/iron oxide composite was prepared to improve the hydrophilicity and biocompatibility of MW-CNTs. Synthesized magnetic MWCNT–starch–iron oxide was used as an adsorbent for removing anionic MO and cationic MB from aqueous solutions. MWCNT–starch–iron oxide exhibits super paramagnetic properties with a saturation magnetization and better adsorption for MO and MB dyes than MWCNT–iron oxide. The specific surface areas of MWCNT/iron oxide and MWCNT–starch-iron oxide were 124.86 and 132.59 m^2/g. However, surface area of the composite was small but adsorption capacity was higher compared to the parent one, the adsorption capacities of MB and MO onto MWCNTs–starch–iron oxide were 93.7 and 135.6 mg/g, respectively,while for MWCNTs–iron oxide were 52.1 and 74.9 mg/g. This is again confirming that ternary composite has the higher removal capacity than the binary composite and as grown CNTs. A novel magnetic composite bio-adsorbent composed of chitosan wrapping magnetic nano sized γ-Fe_2O_3 and multiwalled carbon nanotubes (m-CS/γ-Fe_2O_3/MWCNTs) were prepared for the removal of methyl orange. The adsorption capacity of MO onto m-CS/γ-Fe_2O_3/MWCNTs was 2.2 times higher than m-CS/γ-Fe$_2$O$_3$. The adsorption capacity of MO onto m-CS/γ-Fe$_2$O$_3$/MWCNTs (66.90 mg/g) was also higher than MWCNTs (52.86 mg/g). Kinetics data and adsorption isotherm data were better fitted by pseudo-second-order kinetic model and by Langmuir isotherm, respectively. On the comparison of nature of adsorption of MO onto MWCNTs and m-CS/γ-Fe$_2$O$_3$/MWCNTs, observed that adsorption of MO onto MWCNTs was endothermic. CS is responsible for the exothermic adsorption process because with the increase in the temperature, polymeric network of CS changed/ de shaped, which reduced the porosity of the bio-sorbent and hindered the diffusion of dye molecules at high temperature [72, 73–191].

A fast separation process is obtained by magnetic separation technology. Therefore, the adsorption technique with magnetic separation has aroused wide concern. However, the highest maximum adsorption capacity is not high enough. Polymer shows excellent sewage treatment capacity in the environmental protection. Recently reported that the de colorization efficiency could gets close to 100% in the direct light resistant black G solution by P (AM-DMC)(copolymer of acryl amide and 2-[(methacryloyloxy) ethyl] tri methyl ammonium chloride). Therefore, the integration of CNTs, magnetic materials and polymer could overcome those defects of low adsorption capacity long adsorption time, separation inconvenience and secondary pollution, however, there was few reports about such nano composite. Here in, the first attempt to prepare the MPMWCNT nano composite with the aid of ionic liquid-based polyether and Ferro ferric oxide was made. The structure of the nanocomposite was characterized and the physical properties were investigated in detail. Combined with the characteristics of the three materials, the magnetic nano composite shows excellent properties of short contact time, large adsorption capacity, rapid separation process and no secondary pollution in the adsorption process. Thischapter could provide new adsorption insights in wastewater treatment. Then, the pH value of the final suspensions was adjusted to 10–11 and the redox reaction continued for 30 min with stirring. The MPMWCNT nanocomposite was separated by a permanent magnet and dried under vacuum. The yield of the nanocomposite was about 66.5% [83, 191].

Earlier studies have indicated that magnetic carbon nano composites may show great application potential in magnetic data storage, for magnetic toners in xerography, Ferro fluids, magnetic resonance imaging, and reversible lithium storage. However, there is still lack of a systematic review on the synthesis and application of carbon-based/magnetic nanoparticle hybrid composites. So here we present a short review on the progress made during the past two decades in synthesis and applications of magnetic/carbon nano composites. Synthesis of magnetic carbon nano composites during the last decade, much effort has been devoted to efficient synthetic routes to shape-controlled, highly stable, and well defined magnetic carbon hybrid nano composites. Several popular methods including filling process, template-based synthesis, chemical vapor deposition, hydrothermal/solvo thermal method, pyrolysis procedure, sol-gel process, detonation induced reaction, self-assembly method etc. can be directed at the synthesis of high-quality magnetic carbon nano composites. Examples for the applications of such materials include environmental treatment, microwave absorption, electrochemical engineering, catalysis, information storage, biomedicine and biotechnology [83, 192].

1.7.4   CNT-DOLOMITE, CNT-CELLULOSE AND CNT-GRAPHENE

Recently developed novel adsorbents by inserting MWCNTs into the cavities of dolomites for scavenging of ethidium the foam lines CNTs/dolomite adsorbent. Adsorptions reached equilibrium within 30 min for the cationic dyes, acridine orange, ethidium bromide, and methylene blue while it was about 60 min for the anionic dyes, eosin B and eosin Y. Foam like ternary composite PUF/diatomite/dispersed-MWCNTs, gave the highest capacities for adsorption of these dyes, followed by PUF/agglomerated-MWCNTs, and then PUF/dispersed-MWCNTs. Langmuir adsorption isotherm was best fitted to the equilibrium data. The removal of methylene blue onto natural tentacle type wale gum grafted CNTs/cellulose beads was investigated by some researchers. The maximum adsorption of MB was observed at pH5 and150 min. Adsorption isotherm study revealed the monolayer adsorption at higher concentration and multilayer adsorption at lower concentration. The equilibrium adsorption capacity onto the adsorbent was determined to be 302.1 mg/g at pH 6.0 from Sips model. Pseudo first order kinetics gives the best-fitted results compared to the pseudo second order. From the results, it is evident that carboxylic group on the adsorbent plays the important role for the removal of MB as ionized to COO− at higher pH and bind with MB through electrostatic force [1].Self-assembled cylindrical graphene–MCNT (G–MCNT) hybrid, synthesized by the one pot hydrothermal process was used as adsorbent for the removal of methylene blue in batch process. G–MCNT hybrid showed good performance for the removal of MB from aqueous solution with a maximum adsorption capacity of 81.97 mg/g. The kinetics of adsorption followed the pseudo-second-order kinetic model and equilibrium data were best fitted to Freund lich adsorption isotherm. The adsorption capacity of G–MCNTs is much higher than MCNTs. Therefore, G–CNTs hybrid could be used as an efficient adsorbent for environmental remediation [72, 73].

1.7.5   CNT-CF (PAN) AS ADSORBENT

In order to combine carbon nanotubes with carbon fibers, most studies report the direct synthesis of carbon nanotubes on carbon fibers by CVD (chemical vapor deposition) with special attention paid to the control of CNT length, diameter and density as well as arrangement and anchorage on carbon fibers. The approach, which was mostly investigated, consists in impregnating carbon fibers with a liquid solution of catalyst precursors (nickel, iron or cobalt nitrates, iron chloride) followed by CNT growth from carbonaceous gaseous precursors such as ethylene and methane. Carbon fibers can also be covered by catalyst particles (iron, stainless steel) using sputtering or evaporation techniques or even electrochemical deposition before introducing a gaseous carbon source (methane, acetylene) for CNT growth. In order to improve the CNT growth efficiency on carbon fibers, different modifications have been made such as addition of H_2S in the reactive gas phase, or the development of chemical vapor infiltration (CVI) technique, enabling us to improve the density of CNT and/or their distribution on the fibers. Based on the diffusion of catalyst particles into carbon, as reported in several papers, some studies report the deposition of layers on C fibers playing the role of barrier against catalyst diffusion. Carbon fiber coating with SiO_2-based thick layers through coprecipitation or sol gel methods prior to CNT growth has been investigated, allowing the improvement of CNT density and distribution[4].Growing aligned carbon nanotubes with controlled length and density is of particular importance for multi scale hybrid composites, since such nanotubes are expected to improve electrical and mechanical properties. Injection-CVD techniques for both the deposition of the ceramic sub layer from organ metallic precursors and CNT growth from a hydrocarbon/metallocene precursor mixture.

This process is efficient for the growth of aligned nanotubes on carbon substrates, but also on metal substrates such as stainless steel, palladium, or any metal substrates compatible with the CNT synthesis temperature [185].

The injection-CVD setup is quite similar to the one described by some researchers and has been modified in order to achieve CNT growth on carbon fibers and metal substrates. In particular, the evaporation chamber referred to later as 'the evaporator' was fitted with two identical injectors, derived from standard car engine injectors. A 1 M tetra ethyl ortho silicate (TEOS) solution in toluene was used as precursor for the deposition of silica sub layers. A ferrocene (2.5 wt%) solution in toluene was used for the growth of CNTs. The tank feeding the first injector is filled with the sublayer precursor solution while the tank feeding the second injector is filled with the CNT precursor solution. For the deposition of the silica sub layers, the evaporator temperature is set to 220 °C and the furnace temperature is set to 850 °C. The Ar carrier gas flow rate is adjusted to 1–1 min^{-1} and the pressure is regulated at 100 m bar. The sublayer precursor solution injection rate is 0.96 g min^{-1} TEOS solution was injected into the oven over 23–230 s. Once the sub layer deposition is realized, the evaporator temperature is lowered to 200 °C and the Ar flow rate is adjusted to 3 1 min^{-1}. The injections of the ferrocene/toluene solution with a 0.75 g min^{-1} injection rate over 1–15 min takes place in the same furnace without manipulation of the pretreated fibers. In this experiments, CNT growth was performed over 1–10 min on carbon fibers covered with a SiO$_2$-based layer deposited over the longer duration and was compared to CNT growth on a flat quartz substrate, whereas in CNT growth experiment was performed over 15 min on carbon fibers covered with a SiO$_2$-based layer deposited over variable durations and was compared to CNT growth on a flat quartz substrate. In addition, some experiments were performed on quartz fibers in order to check the influence of the morphology of the substrate on CNT arrangement [186].

Carbon fibers are widely used as reinforcement in composite materials because of their high specific strength and modulus. Such composites have become a dominant material in the aerospace, automotive and sporting goods industries. Current trends toward the development of carbon fibers have been driven in two directions; ultrahigh tensile strength fiber with a fairly high strain to failure (2%), and ultrahigh modulus fiber with high thermal conductivity. Today a number of ultrahigh strength polyacrylonitrile (PAN)-based(more than 6GPa), and ultra high modulus pitch-based(more than 900GPa) carbon fibers have been commercially available. Carbon fibers with exceptionally high thermal conductivity are critical for many thermal control applications in the aerospace and electronics industries. The thermal conductivity of carbon fibers was found to increase asymptotically as the degree of preferred orientation of the crystalline parts in the fiber increases. However, further improvement of thermal conductivity over the existing highly oriented pitch-based carbon fiber while retaining the desired mechanical property has proven to be very challenging. One of the most effective approaches to further increase the thermal conductivity is to graft CNTs on the carbon fibers. CNTs have an extremely high thermal conductivity in the axial direction, and the thermal conductivities of multiwalled CNTs had been reported to be as high as 3000W/m K. The grafting of CNTs on carbon fibers using chemical vapor deposition and electrode position has been reported in the literature. Some researchers reported that the grafting of CNTs improves the mechanical properties and Wei bull modulus of ultra high strength PAN-based and ultra high modulus pitch-based carbon fibers.The effect of grafting CNTs on the thermal conductivity of T1000GB PAN-based and K13D pitch-based carbon fibers were investigated (Fig.10). Recently reported a methods for the self-assembled fabrications of a single suspended amorphous carbon nano wires on a carbon MEMS platform by electro spinning and pyrolysis of PAN and polymers. Here, we explore this technique's potential to fabricate the CNT/PAN composite nano fibers anchored to electrodes and thus, investigate the graphitic and electrical properties of single suspended CNT/carbon composite nano fibers. The conductivity of electro spun carbonized CNT/PAN nano fibers are measured at four different concentrations of MWCNT in the PAN Electro spinning solution. In order to understand the structural changes that are responsible for the increase in conductivity of these nano fibers, micro Raman spectroscopy, X-ray diffraction (XRD) and high-resolution transmission electron microscopy (HRTEM) are used. Results indicate that the crystallinity and electrical conductivity of these composite nano fibers increase with increase in concentration of CNTs [185, 186].

An effective strategy for positioning, integration and interrogation of a single nano fiber requires controlled electro spinning as detailed elsewhere. Researchers also determine the maximum concentration of CNTs in PAN that allows good electro spin ability of the precursor polymer to carbon nano wires. Overall fabrication methodology is a combination of three techniques: (1) photolithography to produce an MEMS structure, (2) self-assembled electro spinning of functionalized CNTs in PAN solution to form nano wires anchored on the MEMS platform, and(3) controlled pyrolysis to obtain carbon composite wires integrated with the underlying carbon MEMS structure. Four different concentrations of CNTs in PAN for electro spinning were prepared by mixing 0.05, 0.1, 0.2 and 0.5wt% CNTs in 8wt% PAN in DMF. The detailed method for solution preparation is described elsewhere. These solutions were electro spun using Dispovan syringe (volume: 2.5 mL and diameter: 0.55 mm) at 13–15kV on the MEMS structures fabricated earlier. The distance between the tip of the jet and the MEMS collector was 10 cm, and the flow rate of the solution was maintained at 1 L/min. Electro spinning was performed for a short period of 5–10 s to limit the number of wires and to obtain single wires suspended between posts. The resulting structures consisting of suspended composite nano wires on polymers were stabilized in air at 250 °C for 1h prior to pyrolyzing the whole structure which is done under N_2 flow (flow rate: 0.2 L/min) at 900 °C for 1h, with a ramp time of 5 °C/min, to yield a monolithic carbon structure having good interfacial contacts and with electrically conducting posts of much greater cross-sectional area than the nano fibers themselves [186].

Carbon fibers used in this study are (i) a low thermal conductivity and ultra high tensile strength PAN-based(T1000GB) carbon fiber and (ii) a high thermal conductivity and ultra high modulus pitch-based(K13D) carbon fiber. Note that as received, both fibers had been subjected to commercial surface treatments and sizing (epoxy compatible sizing). To grow CNTs on the carbon fibers, a Fe $(C_5H_5)_2$(ferrocene) catalysts were applied to the samples fiber bundles using thermal CVD in vacuum. Prior to the application of the catalyst, the carbon fiber bundles were heat treated at 750 °C for an hour in vacuum to remove the sizing. The growth temperature and time for CNTs deposition were selected as 750 °C(T1000GB) and 700 °C(K13D) for 900 s[187].

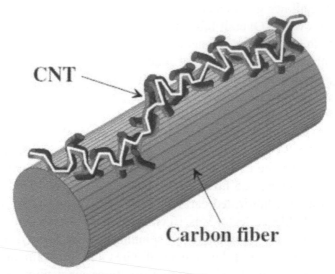

FIGURE 1.3   Schematics model of CNTs-grafted carbon fiber filament.

1.7.6 CNT-AC COMPOSITE

Among various advanced functional materials, electronically conducting polymers (such as polypyrrole (PPy) and poly aniline (PANI)) and metal oxides (such as RuO_2, MnO_2, NiO, and Co_3O_4) are widely used in super capacitors. However, their applications are severely limited by their poor solubility and mechanical brittleness. Anchoring the conductive polymers or metal oxides to cellulose fibers or other textile fibers has inspired the design of their paper or textile based composites which show excellent cycling stability, mechanical flexibility and robustness. However, the textile fibers are usually insulators. Advanced carbon materials, such as CNTs, graphene, ordered mesoporous carbon, carbon aero gels, hierarchical porous carbon, carbide-derived carbon, and their composites/hybrids, have been widely explored for use as super capacitor electrodes. The carbon materials can also be combined with conductive polymers or metal oxides to obtain flexible CNT/PANI, CNF/PANI, graphene/PANI or graphene oxide/PANI, or CNT/CuO composite electrodes with improved electrochemical performance. Very recently, inscrolling nanotube sheets and functional guests into yarns, which contained up to 95wt% of otherwise unspinable particulate or nano fiber powders, has been used to fabricate yarns for use in super capacitors and lithium ion battery materials. However, the large-scale production of inexpensive, flexible electrode remains a great challenge. Super capacitors with AC/CNT nanocomposite electrodes have been shown to exhibit enhanced electrochemical performance compared with CNT-free carbon materials, although the original CNTs were strongly entangled with each other, and acid purification was always required. As a result, those CNTs well dispersed in the electrode were too short to form a self-supporting network. The as-obtained AC/CNT nano composites were still in powder form, and a binder was still needed. Recently it has been shown that vertically aligned CNTs, in which the CNTs with large aspect ratio are well oriented, can be well dispersed into individually long CNTs by a two-step shearing strategy. The as obtained CNT pulp, in which long CNTs have good dispersion in the liquid phase, can be used as a feed-stock for CNT transparent conductive films and Bucky paper. As a result of great efforts to produce mass aligned CNTs, they can be easily produced by radial growth on spheres or intercalated growth in lamellar catalysts. In this contribution, industrially produced aligned CNTs, together with AC powder, were used as raw materials to fabricate flexible electrodes. It is expected that the CNTs will bind AC particles together to give a paper like composite. 90wt%–99wt% of AC was first incorporated into the CNT pulp, and the deposited on a filter to make composites [72–188].

The specific surface area, pore volume and pore size distribution are also the important factors affecting the electrochemical performance of super capacitors. The N_2 adsorption isotherms and pore size distributions of AC powder and CNTs are studied. The quantitative data extracted from the N_2 adsorption isotherms are given. AC presents a type-I isotherm, which is typical for a microporous material. The BET surface area and average pore diameter of AC powder were 1374 m^2/g and 2.35 nm, respectively. After the addition of AC or CNTs, the N_2 adsorption was almost unmodified in the small relative pressure region corresponding to microspores, whereas a noticeable enhancement at $P/P_0 > 0.9$, in the range of mesoporosity, was observed. The pore size distribution of CNTs showed two peaks at about 2.5 and 20 nm. The former may arise from the inner space of CNTs and the latter probably arise from the pores between CNTs. The pore size distribution of AC-CNT-5% also showed two small peaks at about 10 and 25 nm. The former may be generated by the pores in bundles of CNTs formed during the liquid phase process. The latter result from the pores generated by the stacking of CNTs and its size was slightly increased due to the presence of AC particles. The BET surface area and average pore diameter of AC/AB were 1011 m²/g and 2.64 nm. The total pore volume decreased due to the addition of 15wt% AB and 10wt% PTFE. However, the ratio between the mesoporous and micro porous volumes (V_{meso}/V_{micro}) was almost the same. The BET surface area and average pore diameter of AC-CNT-5% were 1223 m²/g and 2.99 nm. The microspore volume of the AC was slightly diminished by the presence of CNTs, where as the mesoporous volume increased from 0.32 to 0.38 cm³/g. The V_{meso}/V_{micro} ratio showed a significant increase [188].

1.7.7   CNT-FIBER COMPOSITES

1.7.7.1   CNT-BASED FIBERS

For many applications, fibrous materials are more suitable than bulk materials. In addition, fiber production techniques tend to be suited for the alignment of nanotubes within the polymer matrix. The researchers observed that the alignment of nanotubes within the composite fiber was improved dramatically by increasing the draw ratio. Many subsequent studies also showed that the mechanical/electrical properties of these composite materials are greatly enhanced by fiber spinning. Apart from the traditional melt spinning methods, composite fibers can also be produced by solution-based processing. The coagulation-spinning method was designed so that the CNTs were attached to each other while they were oriented in a preferential direction by a given flow. Nanotube aggregation was obtained by injecting the CNT dispersion into a rotating aqueous bath of PVA, such that nanotube and PVA dispersions flowed in the same direction at the point of injection. Due to the tendency of the polymer chains to replace surfactant molecules on the graphitic surface, the nanotubes dispersion was destabilized and collapsed to form a fiber. These wet fibers could then be retrieved from the bath, rinsed and dried. Significant rinsing was used to remove both surfactant and PVA. Shear forces during the flow lead to nanotube alignment. These fibers displayed tensile moduli and strength of 9–15GPa and ~150 MPa, respectively. For stretched CNT/DNA/PVA fibers, the tensile moduli and strength were ~ 19GPa and ~ 125 MPa, respectively. In the mean time, the coagulation-spinning method was further optimized by others. They injected the SWCNT dispersion into the center of a co-flowing PVA/water stream in a closed pipe. The wet fiber was then allowed to flow through the pipe before being wound on a rotating mandrel. Flow in more controllable and more uniform conditions in the pipe resulted in more stable fibers. Crucially, wet fibers were not rinsed to remove most of PVA (final SWCNT weight fraction ~ 60%). This resulted in large increases in Young's modulus and strength to 80 and 1.8GPa, respectively. Furthermore, study works have shown that single- and multiwalled CNT based fibers could be drawn at temperatures above the PVA glass transition temperature (~ 180 °C), resulting in improved nanotube alignment and polymer crystallinity. These so-called hot-stretched fibers exhibited values of elastic moduli between 35 and 45GPa and tensile strengths between 1.4 and 1.8GPa, respectively. In an alternative approach, CNT/polymer solutions have been spun into fibers using a dry-jet wet spinning technique. This was achieved by extruding a hot CNT/polymer solution through a cylindrical die. An approximately1–10 cm air gap was retained between the die orifice and the distilled water coagulation bath, which was maintained at room temperature. Significant mechanical property increases were recorded for the composite fibers compared with the control samples with no CNT reinforcement. Another method used recently to form composite-based fibers from solution is electro spinning. This technique involves electrostatic ally driving a jet of polymer solution out of a nozzle onto a metallic counterelectrode. In2003, two groups independently described electro spinning as a method to fabricate CNT-polymer composite fibers. Composite dispersions of CNTs in either PAN or PEO in DMF and ethanol/water, respectively, were initially produced. Electro spinning was carried out using air pressure of 0.1–0.3 kg/cm^2 to force the solution out of a syringe 0.5 mm in diameter at a voltage difference of 15–25kV with respect to the collector. Charging the solvent caused rapid evaporation resulting in the coalescence of the composite into a fiber, which could be collected from the steel plate. Fibers with diameters between 10 nm and 1 m could be produced in this fashion[182].

1.7.7.2   TEXTILE ASSEMBLIES OF CNTS

Some researchers have demonstrated the feasibility in processing of nanotube yarns with high twist spun from nanotube forests, plied nanotube yarn processed from a number of single nanotube yarns with counterdirection twist and 3-D nanotube braids fabricated from 36 to 5-ply yarns. In addition, the plied nanotube yarns and 3-D braids were used as through-the-thickness (Z) yarns in 3-D weaving process. The challenges in making nanotube fibers/yarns with desirable properties, according to recent work, are in achieving the maximum possible alignment of the nanotubes or their bundles within the yarn, increasing the nanotube packing density within the yarn and enhancing the internal bonding among the nanotubes. Following this approach then produced MWCNT single-ply yarns and 5-ply yarns, which were made by over-twisting five single yarns and subsequently allowing them to relax until reaching a torque-balanced state. First, the CVD synthesized MWCNTs are 300 lm long and 10 nm in diameter and they formed about 20 nm diameter bundles in a nanotube forest. Simultaneous draw and twist of the bundles produced 10 lm diameter single yarns. Besides single-ply and 5-ply yarns, many studies have also demonstrated the fabrication of other multiply yarns which have been used in 3-D braiding process as well as Z-yarns in 3-D weaving processes. It is currently possible to produce tens of meters of continuous MWCNT yarns. It is reported that no visible damage to the nanotube yarns is imparted by the braiding process and the 3-D braids are very fine, extremely flexible, hold sufficient load, and are well suited for the use in any other textile formation process, or directly as reinforcement for composites. The reported elastic and strength properties of carbon nanotube composites so far are rather low in comparison with conventional continuous carbon fiber composites. It is believed that the properties can be substantially improved if the processing methods and structures are optimized. Others also studied the electrical conductivity of CNT yarns, 3-D braids and their composites. It is noted that 3-D textile composites, including 3-D woven and 3-D braided materials, combine high in-plane mechanical properties with substantially improved transverse strength, damage tolerance and impact resistance. However, even relatively small volume content of the out-of-plane fibers results in considerable increase of interstitial resin pockets, which contributes to lower in-plane properties. Utilizing fine CNT yarns could dramatically reduce the through the-thickness yarn size while still sufficiently improving the composite transverse properties. Furthermore, they found that the electrical conductivity of 3-D hybrid composites are many times greater than that of commonly produced nano composites made from low volume fraction dispersion of relatively short CNTs in epoxy [183].

1.7.7.3 CARBON NANOTUBE FIBERS

The superb mechanical and physical properties of individual CNTs provide imputes for researchers in developing high-performance continuous fibers basedupon carbon nanotubes. Unlike in the case of carbon fibers, the processing of CNT fibers does not require the cross-linking step of the precursor structures. As summarized results, the leading approaches for the production of CNT fibers are (i) spinning from a lyotropic liquid crystalline suspension of nanotubes, in a wet-spinning process similar to that used for polymeric fibers such as aramids, (ii) spinning from MW-CNTs previously grown on a substrate as "semialigned" carpets and (iii) spinning directly from an aero gel of SWCNTs and MWCNTs as they are formed in a chemical vapor deposition reactor. These methods as well as the twisting of SWCNT film are reviewed. Recently, some researchers reported the properties of epoxy/CNT fiber composites, which are similar to composites reinforced with commercial carbon fibers. The composites were formed by the diffusion of uncured epoxy into an array of aligned fibers of CNTs. The tensile and compressive properties were measured. The results demonstrated significant potential of CNT fiber reinforced composites. The chapter also highlights the issue in defining the cross-section of CNT fibers and other CNT macroassemblies for mechanical property evaluation. Mora et al. noted that the volumetric density of CNT fibers in epoxy composites was found to be much higher than the density of the as-spun fiber obtained from gravimetric and diameter measurements. This difference is related to the volume of free space inside a bundle of fibers. This space is infiltrated by the epoxy and ultimately increases the CNT/polymer interface area. The catalyst particles must be eliminated from the fiber; drawing conditions must be optimized to eliminate entanglements between CNTs; and the fiber needs to be pulled at the rate at which nanotubes are growing so the growth of an individual nanotube is not terminated [183].

1.7.7.4 GEL-SPINNING OF CNT/POLYMER FIBERS

CNTs can act as a nucleation agent for polymer crystallization and as a template for polymer orientation. SWCNTs with their small diameter and long length can act as ideal nucleating agents. Study works suggested that the next-generation carbon fibers will likely be processed not from polyacrylonitrile alone but from its composites with CNTs. Furthermore, continuous carbon fibers with perfect structure, low density, and tensile strength close to the theoretical value may be feasible if processing conditions can be developed such that CNT ends, catalyst particles, voids, and entanglements are eliminated. Such a CNT fiber could have ten times the specific strength of the strongest commercial fiber available today. In the current manufacture process of carbon fibers, polymer solution is extruded directly into a coagulation bath. However, higher strength and modulus PAN and PAN/SWCNT based fibers can be made through gel spinning. In gel spinning, the fiber coming out of the spinneret typically goes into a cold medium where it forms a gel. These gel fibers can be drawn to very high draw ratios. Gel-spun fibers in the gel bath are mostly unoriented and they are drawn anywhere from 10 to 50 times after gelatin. Structure of these fibers is formed during this drawing process. The gel-spinning process, invented around 1980 has been commercially practiced for polyethylene to process high-performance fibers such as Spectra™ and Dyneema™. Although small diameter PAN fibers (10 nm to 1 lm diameter) can be processed by electro spinning, the molecular orientation and hence the resulting tensile modulus achieved is rather low, and processing continuous fiber by electro spinning has been challenging. Others have used island-in-a-sea bi-component geometry along with gel-spinning to process PAN and PAN/CNT fibers to make carbon fibers with effective diameters less than 1 lm. Small-diameter fibers possess high strength and gel-spinning results in high draw ratio and consequently high orientation and modulus [183].

1.7.7.5   ELECTRO SPINNING OF CNT/POLYMER FIBERS

Electro spinning is an electrostatic induced self-assembly process, which has been developed for decades, and a variety of polymeric materials have been electro spun into ultra-fine filaments [22]. Electro spinning of CNT/polymer fibrils is motivated by the idea to align the CNTs in a polymer matrix and produce CNT/polymer nano composites in a continuous manner. The alignment of CNTs enhances the axial mechanical and physical properties of the filaments. Recently researchers have adopted the co-electro spinning technique for processing CNT/PAN (poly acrylonitrile) and GNP (graphite nano platelet)/PAN fibrils. The fluid is contained in a lasso's syringe, which has a capillary tip (spinneret). When the voltage reaches a critical value, the electric field overcomes the surface tension of the suspended polymer and a jet of ultra-fine fibers is produced. As the solvent evaporates, a mesh of nano to micro size fibers is accumulated

On the collection screen the fiber diameter and mesh thickness can be controlled through the variation of the electric field strength, polymer solution concentration and the duration of electro spinning. In the processing of CNT/PAN nanocomposite fibrils, polyacrylonitrile with purified high-pressure CO disproportionate (HiPCO) SWCNTs dispersed in dim ethyl form amide, which is an efficient solvent for SWCNTs, are coelectron spun into fibrils and yarns. CNT-modified surfaces of advanced fibers prepared first time as modified the surface of pitch-based carbon fiber by growing carbon nanotubes directly on carbon fibers using chemical vapor deposition. When embedded in a polymer matrix, the change in length scale of carbon nanotubes relative to carbons fibers results in a multi scale composite, where individual carbon fibers are surrounded by a sheath of nano composite reinforcement. Single-fiber composites have been fabricated to examine the influence of local nanotube reinforcement on load transfer at the fiber/matrix interface. Results of the single-fiber composite tests indicate that the nanocomposite reinforcement improves interfacial load transfer. Selective reinforcement by nanotubes at the fiber/matrix interface likely results in local stiffening of the polymer matrix near the fiber/matrix interface, thus, improving load transfer. The interfacial shear strength of CNT coated carbon fibers in epoxy was studied using the single-fiber composite fragmentation test. Randomly oriented MWCNTs and aligned MWCNTs coated fibers demonstrated 71% and 11% increase in interfacial shear strength over unsized fibers. Sager et al. attributed this increase to the increase in both the adhesion of the matrix to the fiber and the inter phase shear yield strength due to the presence of the nanotubes [183]. Another method to exploit the axial properties of CNTs is to assemble them into a macroscopic fiber, with the tubes aligned parallel with the fiber axis; a strategy similar to that proposed eight decades ago for the development of high-performance polymer fibers. Carbon nanotube fibers can be produced by drawing from an array of vertically aligned CNTs, by wet-spinning from a liquid crystalline suspension of CNTs or they can be spun directly from the reactor by drawing them out of the hot-zone during CNT growth by CVD. Considerable attention has been devoted to optimizing the structure of CNT fibers at the different stages of their production:

controlling the synthesis of specific nanotubes, the assembly of CNTs into a fully dense fiber and using post-spin treatments to obtain specific properties. However, the integration of CNT fibers into composites and the properties of these composites have received comparatively less attention, in spite of these aspects being fundamental for many potential applications of this new high-performance fiber. A previous study on the mechanical properties of CNT fiber/epoxy composites showed effective reinforcement of the thermoses matrix both in tension and compression without the need for additional treatments on the CNT fiber. Large increases in stiffness, energy absorption, tensile strength and compressive yield stress were observed for composite with fiber mass fractions in the range 10–30%. CNT fibers have an unusual yarn-like structure with an accessible surface area several orders of magnitude higher than that of a traditional fiber. The free space between bundles in the CNT fiber is able to take up non cross-linked resins by capillary action, which wick into the fiber and appear to fill the observable free space. As a consequence, the composite develops a hierarchical structure, with each CNT fiber, being an infiltrated composite itself. Measurements of CTE of CNT fiber composites show good stress transfer between matrix and fiber due to the good adhesion of the thermo set to the porous fibers aided by a significant level of structural 'keying.' The adding polymer into the CNT fibers does not disrupt the CNT bundle network; hence the electrical conductivity of the fiber is largely preserved and the electrical conductivity of the composite is increased. On the other hand, the thermal conductivity of the CNT fiber composite increases more rapidly than before with added fiber loading up to the maximum used here of 38%. Others suggest that the infiltration of the polymer into the fiber improves the thermal coupling between the nanotube bundles by filling the spaces between them, which is a characteristic of the as-spun condition. Addition of CNT fiber to the matrix produces an effective increase in composite thermal conductivity of 157W/m K per unit fiber mass fraction. Results show easy integration of CNT fibers into axially aligned composites and rather effective exploitation of the electrical, thermal and mechanical axial properties of the CNT fibers thanks to the infiltration of polymer into them [184] (Fig. 1.4).

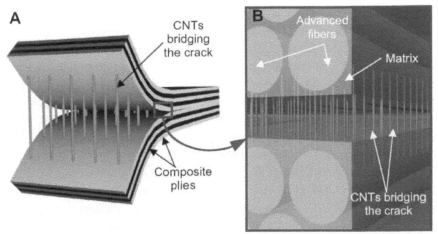

FIGURE 1.4 Illustrations of the ideal hybrid inter laminar architecture: (A) CNTs placed in between two plies of a laminated composite and (B) close-up of the crack, showing CNTs bridging the crack between the two plies. Illustrations are not to scale.

1.7.8 CNT-PANI COMPOSITES AS ADSORBENT

This study confirms that DP/MWCNTs are a potential adsorbent for the removal of Cr(VI) from aqueous solution. The functional groups present on the surface of adsorbent mainly, amine or imines groups of poly aniline and exposed surface and internal porous structure sites were mainly responsible for the adsorption of Cr(VI). The optimum Cr(VI) adsorption was found at pH 2. The equilibrium isotherm and kinetic results suggested that the adsorption was heterogeneous, multilayer and controlled through a boundary layer effect. The higher value of enthalpy change (DH_) and incomplete desorption suggested that chemisorptions was main mechanism involved in the adsorption of Cr(VI)[180].

Due to its outstanding stability in acidic solution, organic magnetic materials have attracted considerable attention and are viewed as a promising substitute for metal and metal oxide based magnetic materials. Among them, PANI is one of the most attractive magnetic polymers because of its excellent chemical stability, easy synthesis and excellent magnetic property. Because the large amount of amine and imines functional groups of PANI are expected to have strong affinity with metal ions, PANI can enrich and remove heavy metal contaminants from aqueous solutions easily and effectively. Considering the promising magnetic property of PANI and excellent adsorption capacity of MWCNTs, PANI/MWCNTs magnetic composites might be an attractive material in the removal of heavy metal contaminants from large volumes of aqueous solutions. Comparing to conventional chemical methods, plasma technique has many advantages in surface modification of various materials. Some researchers also modified single walled carbon nanotubes with PANI by using plasma induced polymerization technique and studied its electrical properties. Here in, MWCNTs were modified with PANI by using plasma induced polymerization technique to synthesize PANI/MWCNTs magnetic composites. The prepared PANI/MWCNTs magnetic composites were characterized by ultraviolet–visible (UV-vis) spectra photometry, X-ray photoelectron spectroscopy (XPS), thermo-gravimetric analysis differential thermal analysis TGA–DTA), field-emission scanning electron microscopy (FE-SEM), and vibrating sample magnetometer (VSM). The prepared PANI/MWCNTs magnetic composites were applied to adsorb and to enrich Pb(II) from aqueous solutions to evaluate its application in the removal of heavy metal contaminants from large volumes of aqueous solutions in environmental pollution cleanup. In the application of PANI/MWCNTs to remove Pb(II) ions in environmental pollution management, the pH value of aqueous solutions is also crucial because the adsorption of metal ions on solid particles is generally affected by the pH values. Moreover, the amine and imines functional groups of PANI and the oxygen containing functional groups of MWCNTs on PANI/MWCNT surfaces are prorogated at low pH. The functional groups of PANI/MWCNTs are prorogated easily at low pH; causing PANI/MWCNTs carry positive charges. Moreover, at low pH, the functional groups on PANI/MWCNT surfaces are competitively bound by the protons in aqueous solutions, which can restrict the enrichment of Pb(II) onto PANI/MWCNTs. The protonation of functional groups on PANI/MWCNT surfaces decreases with increasing pH, which results in the less positively charged PANI/MWCNTs. It is reasonable that the competitive binding for the functional groups on PANI/MWCNT surfaces between protons and Pb(II) will decrease. The removal of Pb(II) on MWCNTs and PANI/MWCNTs as a function of pH is dependent on pH values. The effect of ionic strength on the sorption of Pb(II) to PANI/MWCNTs is also important because of the presence of different cat ions and anions in the environment and the salt concentration in wastewater may be different for different sites. PANI/MWCNT magnetic composites were synthesized by using plasma induced polymerization technique. The analysis results of

UV-vis spectra, XPS, TGA, and FE-SEM characterizations indicate that PANI has been modified onto MWCNTs. PANI/MWCNT magnetic composites have very high adsorption capacities in the removal of Pb(II) ions from large volumes of aqueous solutions, and PANI/MWCNTs magnetic composites can be separated and recovered from solution by simple magnetic separation. The results of this research highlight the potential application of the magnetic composites of PANI/MWCNTs for heavy metal contaminants cleanup in the natural environment [181].

1.7.9 CNT-ALGINATE COMPOSITES AS ADSORBENT

Widespread usage of CNTs will cause increasing emissions to the water environment and result in human contact risk to CNTs. Another adverse factor currently restricting the application of CNTs to environmental protection/remediation is their high costs. These difficulties encountered in environmental applications with alginate and CNTs can be overcome by formation of calcium alginate (CA)/MWCNTs composite fiber. As defined in our laboratory, CA/MWCNT composite fiber was prepared using CaCl2 as cross-linking agent by wet spinning. The composite fibers not only make full use of the heavy metals and dyes adsorption properties of alginate and MWCNTs, but also prevent MWCNTs from breaking off the composites to cause second micro pollution to water. Such technique is also a practical approach to overcome the high cost difficulty encountered in the use of CNTs for environmental remediation. In the recent work, the adsorption properties for MB and MO organic dyes on CA/MWCNT composite fiber were characterized by using batch adsorption method. For preparation of SA/MWCNT dispersions, usually SA was dissolved in distilled water to produce a viscous solution with the concentration of 3wt% SA after mechanical agitation for 4h at ambient temperature and continue stirring for 30 min at 50 °C. Then different amount of MWCNTs were added into SA solution under constant stirring for 30 min at room temperature and then ultrasonic 15 min to achieve homogeneous dispersion. Preparation of CA/MWCNT composite fibers spinning dope was prepared for different concentrations of nanotube and SA. A narrow jet of spinning solution was injected through a 0.5 mm diameter needle into a coagulation bath containing 5wt% aqueous solution of $CaCl_2$, and then collected on a spindle outside the bath, which was rotated. The coagulated fibers were then washed several times with de-ionized water and then dried in air under tension. The CA/MWCNT bio composite fibers obtained had a linear-density of about 50dtex. A novel material, the CA/MWCNTs composite fiber as an effective bio-adsorbent for ionic dyes removal has been prepared. The adsorbent toward ionic dyes has higher adsorption capacity due to the introduction of MWCNTs and the large specific area of fiber form of the composite. Batch adsorption experiments showed that the adsorption process followed second-order kinetic model. The results suggested that the initial pH value is one of the most important factors that affect the adsorption capacity of the dyes onto CA/MWCNTs. The desorption experiments showed the percentage of the desorption were found to be 79.7% and 80.2% for MB at pH 2.0 and MO at pH 13.0, respectively, which were corresponding to the adsorption experiments. The CA/MWCNT composite fiber can be used as environment friendly bio adsorbent for the removal of ionic dyes from aqueous solution due to the efficient and fast adsorption process [178].

One effective method to resolve the second pollution caused by CNTs is to search for suitable supporters to immobilize CNTs for preparing macroscopic CNTs composites in order to make full use of the current microsized CNTs. Alginate, the salt of alginic acid, has hydrophilicity, biocompatibility, non toxicity, exceptional formability and is a linear chain structure of guluronic acid (G) residues arranged in a block wise fashion. It has a high affinity and binding capacity for metal ions and has already been widely used as heavy metal adsorbent in environmental protection. Some researchers reported that the adsorption isotherms of UO_2^{2+} adsorbed by calcium alginate beads exhibited Langmuir behavior and its maximum molecular capacity reached 400 mg/g for UO_2^{2+} at 25 °C. Others investigated Cu^{2+} removal capability of calcium alginate encapsulated magnetic sorbent and found that the maximum sorption capacity of the material can reach 60 mg/g at pH of 5.0 and temperature of 20 °Cand too prepared composite gels of calcium alginate containing iminodiacetic type resin. The biosorption capacity of the composite gels is higher than that of simple alginate gels and increases with increasing the amounts of resin enclosed in the composite. The adsorbents were characterized using SEM micrograph, BET surface analysis, FTIR spectra and Boehm titration method. The CA and CNTs/CA were deposited on a brass hold and sputtered with thin coat of gold under vacuum, their morphology and surface CNTs. The results suggest that the higher surface area and pore volume of

CNTs maybe act as micro channels in matrix and benefit for the improvement of surface area of the CNTs/CA composites. In summary, an efficient adsorbent of CNTs/CA with exceptional Cu(II) adsorption capability was prepared. Equilibrium data were fitted to Langmuir and Freund lich isotherms. Based on Langmuir isotherm, the maximum monolayer adsorption capacity for CNTs/CA is 84.88 mg/g. The new type of adsorbent of CNTs/CA can resolve the micropollution problem caused by nano-sized CNTs through immobilizing them by CA and it will promote the practical applications of CNTs and their composites in environmental protection [179].

1.8 CARBON NANOFIBER (CNF)

Carbon nano fibers (diameter range, 3–100 nm) have been known for a long time as a nuisance that often emerges during catalytic conversion of carbon-containing gases. The recent outburst of interest in these graphitic materials originates from their potential for unique applications as well as their chemical similarity to fullerenes and carbon nanotubes. In this reviews, focused on the growth of nano fibers using metallic particles as a catalyst to precipitate the graphitic carbon. First, summarized some of the earlier literatures that has contributed greatly to understand the nucleation and growths of carbon 'nano fibers and nanotubes. Thereafter, described in detail recent progress to control the fiber surface structure, texture, and growth into mechanically strong agglomerates. It is argued that carbon nano fibers are unique high surface area materials (200 mL/g) that can expose exclusively either basal graphite planes or edge planes. It is shown that the graphite surface structure and the lyophilicity play a crucial role during metal emplacement and catalytic use in liquid phase catalysis. An article by Iijima that showed that carbon nanotubes are formed during arc-discharge synthesis of C_{60} and other fullerenes also triggered an outburst of the interest in carbon nano fibers and nanotubes. These nanotubes may be even single walled, whereas low temperature, catalytically grown tubes are multiwalled. It has been realized that the fullerene type materials and the carbon nano fibers known from catalysis are relatives and this broadens the scope of knowledge and of applications. It has been realized, however, that arc-discharge and laser-ablation methods lead to mixtures of carbon materials and thus to a cumber some purification to obtain nano fibers or nanotubes [88].

From an application point of view, some of best application of carbon nano fibers include: Carbon nano fibers as catalyst support materials, Carbon nano fiber based electrochemical biosensors, CNF-based oxidase biosensors, CNF-based immunosensor and cell sensor and hydrogen storage. The overall economics are affected by the fiber yield, the feedstock used, the rate of growth, and the reactor technology [88–92]. The growth of parallel fibers using iron as the catalyst has been studied in detail by HR-TEM. It is noted that the graphite layers grow at an angle iron surface, thus leading to parallel fibers. The diameter of the fibers can be varied by variation of the metal particle size. If we want to vary the fiber diameter for a macroscopic sample, however, we need a narrow metal particle size distribution. In general, one can say that the fibers do not contain microspores and that the surface area can range from 10 to 200 m^2/g and the mesopore volume ranges between 0.50 and 2.0 mL/g. Note that these pore-volume data are obtained with fibers as grown, specific treatments in the liquid phase can be applied to largely reduce the pore volume and to obtain much denser and compact fiber structures. Compared to the large volume of literature on the mechanism of growth, the studies on the macroscopic, mechanical properties of bodies consisting of agglomerates of carbon nano fibers have been limited in number. Give a useful description of the tertiary structures that can be obtained (i.e., bird nests, Neponset, and combed yarn). In general, porous bodies of carbon nano fibers are grown from porous supported metal catalyst bodies. Some others in the size range of micrometer to millimeter. As carbon precursors, PAN and pitches were frequently used, probably because both of them are also used in the production of commercial carbon fibers. In addition, PVA, polyimides (PIs), polybenzimidazol (PBI) poly (vinylidene fluoride) (PVDF), phenolic resin and lignin were used. In order to convert electro spun polymer nano fibers to carbon nano fibers, carbonization process at around 1000 °C has to be applied. In principle, any polymer with a carbon backbone can potentially be used as a precursor. For the carbon precursors, such as PAN and pitches, so-called stabilization process before carbonization is essential to keep fibrous morphology, of which the fundamental reaction is oxidation to change resultant carbons difficult to be graphitized at high temperatures as 2500 °C [89, 90].

Carbon is an important support material in heterogeneous catalysis, in particular for liquid-phase catalysis. A metal support interaction between Ru and C was suggested as a possible explanation for these very interesting observations. More recent work focuses on the use of platelet type fibers exposing exclusively graphite edge sites. Using a phosphorus-based treatment, preferential blocking of so-called armchair faces occurs. Deposition of nickel onto the thus modified CNF enabled one to conclude that the nickel particles active for hydrogenation of light alkenes reside on the zigzag faces. More characterization work is needed to substantiate these interesting claims. Others have carried out, by far, the most extensive work on CNF as carbon support material. A driving force for exploring CNF supports was related to the replacement of active carbon as support for liquid phase catalysis. For the CNF support, no shift of the PSD is apparent, whereas with an AC support, severe attrition is apparent. They compare the PSD after ultrasonic treatment of the CNF and of the AC support. Clearly, AC displays a much broader PSD with, moreover, a significant number of fines. Nanocomposite electrodes made of carbon nano fibers and paraffin wax were characterized and investigated as novel substrates for metal deposition and stripping processes. Since CNFs have a much larger functionalized surface area compared to that of CNTs, the surface active groups to volume ratio of these materials is much larger than that of the glassy-like surface of CNTs. This property, combined with the fact that the number and type of functional groups on the outer surface of CNFs can be well controlled, is expected to allow for the selective immobilization and stabilization of bio molecules such as proteins, enzymes, and DNA. Additionally, the high conductivity of CNFs seems to be ideal for the electrochemical transduction. Therefore, these nanomaterials can be used as scaffolds for the construction of electrochemical biosensors [64–68].

Compared with conventional ELISA-based immunoassays, immuno sensors are of great interest because of their potential utility as specific, simple, label-free and direct detection techniques and the reduction in size, cost and time of analysis. Due to its large functionalized surface area and high surface active groups-to-volume ratio. Hydrogen storage is an essential prerequisite for the widespread deployment of fuel cells, particularly in transport. Huge hydrogen storage capacities, up to 67%, were reported. Unfortunately such astonishing values could not confirmed by other research teams worldwide. Volumetric and gravimetric hydrogen density for different storages methods some reviews provides basics of hydrogen storage on carbon materials, the types of carbon materials with potential for hydrogen storage, the measured hydrogen storage capacities of these materials, andbased on calculations, an approximation of the theoretical achievable hydrogen storage capacity of carbon materials [91, 92].

1.9   ACTIVATED CARBON NANOFIBERS (ACNF): PROPERTIES AND APPLICATION

Electro spinning, a simple approach to make very fine fibers ranging from nano to micro scales, is attracting more attention due to the high porosity and high surface area to volume ratio of electro spun membranes. These properties contribute to potential applications of electro-spun membranes in carbon and graphitic nano fiber manufacturing, tissue scaffolding, drug delivery systems, filtration and reinforced nano composites. The researchers also tried poly (amic acid) as a precursor to make activated carbon nano fibers. These studies employed physical activation to produce pores in precursor fibers. Compared to physical activation, the chemical activation process has important advantages, including low heat treatment temperature, short period of processing time, large surface area and high carbon yield; however, there has been no work reported for chemically activated carbon nano fibers from electro spun PAN. The effects of electro spinning variables, such as applied voltage, pump flow rate and distance between the needle tip and collector, on the resulting nano fiber diameters were studied. Mechanical properties, such as tensile, tear and burst strength, of electro spun PAN nonwoven membranes were measured and the quantitative relationships between membrane thickness and these properties were established [93–95]. Other physical properties, such as air permeability, inter fiber pore size and porosity, were also studied. Activated carbon nano fibers were produced from electro spun PAN by chemical activation with potassium hydroxide (KOH) as the activating agent. They were characterized by morphology, Fourier transform infrared spectroscopy (FTIR) Brunauer-Emmett-Teller (BET) surface area total pore volume and pore size distribution. There are two processes for manufacturing the carbon nano fiber (CNF), namely, the vapor-grown approach and the polymer spinning approach. The ACNF is the physically or chemically activated CNF, which have been, in many investigations, practically applied in electric double layer capacitors, organic vapor recovery, catalyst support, hydrogen storage, and so on. In practice, the physical activation method involves carbonizing the carbon precursors at high temperatures and then activating CNF in an oxidizing atmosphere such as carbon dioxide or steam [96, 97].

The chemical activation method involves chemically activating agents such as alkali, alkaline earth metals, some bases such as KOH and sodium hydroxide, zinc chloride, and phosphoric acid (H_3PO_4). In essence, most chemical activation on CNF used KOH to get highly porous structure and higher specific surface area. Unfortunately, large amount of solvent were needed to prepare the polymer solution for electro spinning and polymer blend, causing serious environmental problem thereafter. A series of porous amorphous ACNF were studied. Utilizing the core/shell microspheres, which were made of various polymer blends with solvents. In their approach, the phenol formaldehyde-derived CNF were chemically activated by the alkaline hydroxides and the thus-prepared ACNF were applied as super-capacitor electrodes and hydrogen storage materials [98–100].

In a continuous effort, some researchers proceed to investigate and compare the various chemical activation treatments on the CNF thus prepared, with particular emphasis on the qualitative description and quantitative estimation on the surface topology by AFM, and their relation to the microstructure of ACNF. PAN fiber following the spinning process by several ways such as modification through coating, impregnation with chemicals (catalytic modification) and drawing/stretching with plasticizer. The post spinning modifications indirectly affect and ease the stabilization in several ways such as reducing the activation energy of cyclization, decreasing the stabilization exothermic, increasing the speed of cyclization reaction, and also improving the orientation of molecular chains in the fibers. One of the well-known posts spinning treatment for PAN fiber precursor is modification through coatings. The PAN fibers are coated with oxidation resistant resins such as lubricant (finishing oil), antistatic agents, andemulsifiers, which are basically used as, spin finish on the precursor fiber. Coating with certain resins also acts in the same manner as the co monomer in reducing the cyclization exothermic thus improving the mechanical properties of the resulting carbon fibers. Due to their excellent lubricating properties, silicone based compounds are mostly used as the coating material for PAN precursor fibers. Tensile load and tear strength of electro spun PAN membranes increased with thickness, accompanied with a decrease in air permeability; however, burst strength was not significantly influenced by the thickness [101–103].

Electro spun PAN Nano fiber membranes were stabilized in air and then activated at 800 °C with KOH as the activating agent to make activated carbon nano fibers. Stabilized PAN membranes showed different breaking behaviors from those before stabilization. The activation process generated microspores which contributed to a large surface area of 936.2 m^2/g and a microspore volume of 0.59 cc/g. Pore size distributions of electro spun PAN and activated carbon nano fibers were analyzed based on the Dubinin-Astakhov equation and the generalized Halsey equation. The results showed that activated carbon nano fibers had many more microspores than electro spun PAN, increasing their potential applications in adsorption. Based on a novel solvent-free co extrusion and melt-spinning of polypropylene/(phenol formaldehyde polyethylene)based core/sheath polymer blends, a series of ACNFs have been prepared and their morphological and microstructure characteristics analyzed by scanning electron microscopy, atomic force microscopy (AFM), Raman spectroscopy, and X-ray diffractometry with particular emphasis on the qualitative and quantitative AFM analysis. Post spinning treatment of the current commercial PAN fiber based on the author's knowledge, reports on the post spinning modification process of the current commercial fiber are still lacking in the carbon fiber manufacturers' product data sheet, which allows us to assume that the current commercial carbon fibers still do not take full advantage of any of these treatments yet. During stabilization and carbonization of polymer nano fibers, they showed significant weight loss and shrinkage, resulting in the decrease of fiber diameter [104–107].

From an application point of view, some of best application of carbon nano fibers include: ACNF as anodes in lithium-ion battery, Organic removal from waste water using, ACNF as cathode catalyst or as anodes for microbial fuel cells (MFC), Electrochemical properties of ACNF as an electrode for super capacitors, Adsorption of some toxic industrial solutions and air pollutants on ACNF [108–120]. ACNFs with large surface areas and small pores were prepared by electro spinning and subsequent thermal and chemical treatments. These activated CNFs were examined as anodes for lithium-ion batteries (UBs) without adding any nonactive material. Their electrochemical batteries show improved lithium-ion storage capability and better cyclic stability compared with in activated counter parts. The development of high-performance rechargeable lithium ion batteries (LIBs) for efficient energy storage has become one of the components in today's information rich mobile society [114].

MFC technologies are an emerging approach to wastewater treatment. MFCs are capable of recovering the potential energy present in wastewater and converting it directly into electricity. Using MFCs may help offset wastewater treatment plant operating costs and make advanced wastewater treatment more affordable for both developing and industrialized nations. In spite of the promise of MFCs, their use is limited by low power generation efficiency and high cost. Some researchers conclude that the biggest challenge for MFC power output lies in reactor design combining high surface area anodes with low ohmic resistances and low cathode potential losses. Power density limitations are typically addressed by the use of better-suited anodes, use of mediator's modification to solution chemistry or changes to the overall system design. Employing a suitable anode, however, is critical since it is the site of electron generation. An appropriately designed anode is characterized by good conductivity, high specific surface area, biocompatibility and chemical stability. Anodes currently in use are often made of carbon and/or graphite. Some of these anodes include but are not limited to: graphite plates/rods/felt, carbon fiber/cloth/ foam/paper and reticulated vitreous carbon (RVC). Carbon paper, cloth and foams are among the most commonly used anodes and their use in MFCs has been widely reported employ ACNF as the novel anode material in MFC systems. Compared with other activated carbon materials, the unique features of ACNF are its mesoporous structure, excellent porous interconnectivity, and high bio available surface area for bio film growth and electron transfer [115].

Among the diverse carbonaceous adsorbents, ACF is considered to be the most promisingdue to their abundant microspores large surface area, and excellent adsorption capacity. Therefore, the investigation of formaldehyde adsorption has been steadily conducted using ACFs. However, most of precedent works have generally concerned about removal of concentrated formaldehyde in aqueous solution (formalin, and thus there was limited information whether these materials could be used in the practical application, because the concentration of formaldehyde in indoor environment was generally very low (below 1ppm). The nitrogen containing functional groups in ACF played an important role in increasing formaldehyde adsorption ability, as also described elsewhere. However, the PAN-based ACFs still have problems in a practical application, because the adsorption capability is drastically reduced under humid condition [116–120].

1.10   POREFORMATION IN CARBON MATERIALS

About pore formation in carbon materials, this is accepted that all carbon materials, except highly oriented graphite, contain pores, because they are poly crystalline and result from thermal decomposition of organic precursors. During their pyrolysis is and carbonization, a large amount of decomposition gases is formed over a wide range of temperatures, the profile of which depends strongly on the precursors. Since the gas evolution behavior from organic precursors is strongly dependent on the heating conditions, such as heating rate, pressure, etc., the pores in carbon materials are scattered over a wide range of sizes and shapes. These pores may be classified as shown in Table 1.3 [1–3].

TABLE 1.3   Classification of Pore Formation in Carbon Materials

Based on their origin		Based on their size			Based on their state
Intra particle pores	Intrinsic intra particle pores	Micro-spores	< 2 nm		Open pores
	Extrinsic intra particle pores				
Inter particle pores	Rigid inter particle pores	Meso-spores	2~50 nm	Ultra micro-spores <0.7 nm	Closed pores (Latent pores)
	Flexible inter particle pores				
		Macro-spores	> 50 nm	Super micro-spores 0.7–2 nm	

Table 1.3 shows that based on their origin, the pores can be categorized into two classes, intra particle and inter particle pores. The intra particle pores are further classified into two, intrinsic and extrinsic intra particle pores. The former class owes its origin to the crystal structure, that in most activated carbons, large amounts of pores of various sizes in the nanometer range are formed because of the random orientation of crystallites; these are rigid inter particle pores. A classification of pores based on pore sizes was proposed by the International Union for Pure and Applied Chemistry

(IUPAC). As illustrated in Table 1.3 pores are usually classified into three classes: macrospores (>50 nm), mesopore (2–50 nm) and microspores (<2 nm). Microspores can be further divided into super microspores (with a size of 0.7–2 nm) and ultra microspores (<0.7 nm in size).Since nano technology attracted the attention of many scientists recently the pore structure has been required to be controlled closely. When scientists wanted to express that they are controlling pores in the nanometer scale, some of them preferred to call the smallest pores nano-sized pores, instead of micro/mesopore [1, 2].

Pores can also be classified on the basis of their state, either open or closed. In order to identify the pores by gas adsorption (a method which has frequently been used for activated carbons), they must be exposed to adsorb ate gas. If some pores are too small to accept gas molecules they cannot be recognized as pores by adsorb ate gas molecules. These pores are called latent pores and include closed pores. Closed pores are not necessarily in small size. Pores in carbon materials have been identified by different techniques depending mostly on their sizes. Pores with nano-meter sizes, that is, microspores and mesopore, are identified by the analysis of gas adsorption isotherms, mostly of nitrogen gas at 77 K [41–46].

The basically theories, equipments, measurement practices, analysis procedures and many results obtained by gas adsorption have been reviewed in different publications. For macrospores, mercury porosimetry has been frequently applied. Identification of intrinsic pores, the interlayer space between hexagonal carbon layers in the case of carbon materials, can be carried out by XRD.Recently, direct observation of extrinsic pores on the surface of carbon materials has been reported using microscopy techniques coupled with image processing techniques, namely STM and AFM and TEM for microspores and mesopore, and scanning electron microscopy (SEM) and optical microscopy for macro pores [1–3].

The most important Pore Characterization methods are include: STM, AFM, TEM, Gas adsorption, Calorimetric methods, Small-angle X-ray scattering (SAXS), Small-angle Neutron Scattering (SANS), Positron Annihilation Lifetime Spectroscopy (PALS), SEM, Optical Microscopy, Mercury Porosimetry and Molecular Resolution Porosimetry. Adsorption from solution using macromolecules has been applied to macro pore analysis, but we still need more examinations. It is difficult to compare one adsorption isotherm with another, but determination of the deviation from the linearity using a standard adsorption isotherm is accurate. The plot constructed with the aid of standard data is called a comparison plot. The representative comparison plots are the t and alpha plots [41, 42].

The molecular adsorption isotherm on non-porous solids, which can be well described by the BET theory. The deviation from the linearity of the t plot gives information on the sort of pores, the average pore size, the surface area, and the pore volume. However, the t plot analysis has the limited applicability to the microporous system due to the absence of explicit monolayer adsorption. The construction of the alpha plot does not need the monolayer capacity, so that it is applicable to micro porous solids. The straight line passing the origin guarantees multilayer adsorption, that is absence of meso-and/or micropores; the deviation leads to valuable information on the pore structures. A nonporous solid has a single line passing the origin, while the line from the origin for the alpha plot of the mesoporous system bends. The slope of the straight line through the origin and the line at the high region gives the total and mesoporous surface areas, respectively. The type of the alpha plot suggests the presence of ultra micro pores and/or super micro pores. Detailed analysis results will be shown for the micro pore analysis [43–46].

In the discussion of the mesopore shape, the contact angle is assumed to be zero (uniform adsorbed film formation). The lower hysteresis loop of the same adsorb ate encloses at a common relative pressure depending to the stability of the adsorbed layer regardless of the different adsorbents due to the so called tensile strength effect. This tensile strength effect is not sufficiently considered for analysis of meso pore structures. The Kelvin equation provides the relationship between the pore radius and the amount of adsorption at a relative pressure. Many researchers developed a method for the calculation of the pore size distribution on the basis of the Kelvin equation with a correction term for the thickness of the multilayer adsorbed film. They so called BJH (Barret-Joyer-Halenda) and DH (Dollimore-Heal) methods have been widely used for such calculations, as mathematical details are shown in other articles are only the simple FORTRAN program for the DH method (this program can be easily used for the analysis of the mesopore size distribution).The thickness correction is done by the Dollimore-Heal equation. One can calculate the mesopore size distribution for cylindrical or slit shaped mesopore with this program. Therefore, the adsorption branch provides more reliable results. However, the adsorption branch gives a wide distribution compared to the desorption branch due to gradual uptake. Theoretical studies on these points are still done [133–162].

The pore size distribution from the Kelvin equation should be limited to mesopore due to the ambiguity of the meniscus in the microporous region. It is well known that the presence of microspores is essential for the adsorption of small gas molecules on activated carbons. However, when the adsorbate is polymer, dye or vitamin, only mesopore allow the adsorption of such giant molecules and can keep even bacteria. The importance of mesopore has been pointed out not only for the giant molecule adsorption, but also for the performance of new applications such as electric double layer capacitors. Thus, the design and control of mesoporosity is very desirable both for the improvement of performance of activated carbon and for the development of its new application fields [1–3].

Important parameters that greatly affect the adsorption performance of a porous carbonaceous adsorbent are porosity and pore structure. Consequently, the determination of PSD of carbon nano structures adsorbents is of particular interest. For this purpose, various methods have been proposed to study the structure of porous adsorbents. A direct but cumbersome experimental technique for the determination of PSD is to measure the saturated amount of adsorbed probe molecules, which have different dimensions. However, there is uncertainty about this method because of networking effects of some adsorbents including activated carbons and carbon nano structures. Other experimental techniques that usually implement for characterizing the pore structure of porous materials are mercury porosimetry, XRD or SAXS, and immersion calorimetry. A large number of simple and sophisticated models have been presented to obtain a realistic estimation of PSD of porous adsorbents. Relatively simple but restricted applicable methods such as Barret, Joyner and Halenda (BJH), Dollimore and Heal (DH), Mikhail et al. (MP), Horvath and Kawazoe (HK), Jaroniec and Choma (JC), Wojsz and Rozwadowski (WR), Kruk-Jaroniec-Sayari (KJS), and Nguyen and Do (ND) were presented from 1951 to 1999 by various researchers for the prediction of PSD from the adsorption isotherms [133, 139, 162].

For example, the BJH method, which is usually recommended, for mesoporous materials is in error even in large pores with dimension of 20 nm. The main criticism of the MP method, in addition to the uncertainty regarding the multilayer adsorption mechanism in microspores is that we should have a judicious choice of the reference isotherm. HK model was developed for calculating microspore size distribution of slit-shaped pore; however, the HK method suffers from the idealization of the microspore filling process. Extension of this theory for cylindrical and spherical pores was made by Saito and Foley and Cheng and Ralph. By applying some modifications on the HK theory, some improved models for calculating PSD of porous adsorbents have been presented. Gauden et al. extended the Nguyen and Do method for the determination of the bimodal PSD of various carbonaceous materials from a variety of synthetic and experimental data. The pore range of applicability of this model besides other limitations of ND method is its main constraint. In 1985, Bunke and Gelbin determined the PSD of activated carbons based on liquid chromatography (LC). Choice of suitable solvent and pore range of applicability of this method are two main problems that restrict its general applicability. More sophisticated methods such as MD, Monte Carlo simulation, Grand Canonical Monte Carlo simulations (GCMS), and density functional theory (DFT) are theoretically capable of describing adsorption in the pore system. The advantages of these methods are that they can apply on wide range of pores width. But, they are relatively complicated and provide accurate PSD estimation based on just some adsorbates with specified shapes [133, 140, 145, 164, 165].

1.11 NOVELMETHODES FOR CONTROLLING OF PORE STRUCTURE IN CARBON MATERIALS

1.11.1 ACTIVATION

Activation processes are often classified into two, gas activation and chemical activation. Different oxidizing gases, such as air, CO_2, and water vapor, were used for gas activation. For chemical activation, $ZnCl_2$ and KOH were used as an activation agent. Recently, a great success to get a high surface area reaching to approximately 3600 m^2 g^{-1} was obtained by using KOH for activation and applied for the preparation of some of activated carbons. During activation process, the creation of microspores is the most important. In most of carbon materials, however, macrospores and mesopore coexisted with microspores. In other words, macrospores and mesopore had to be formed during the activation process in order to develop a large number of microspores. On adsorption and desorption procedure, however, these pores play an important role as diffusion pathways for adsorbates to microspores. For the carbons prepared from biomasses, such as coconut shells and wood chips, many macrospores are already formed during carbonization as a memory of cell structure of the original biomasses, which seems to make microspore development by activation easier. In activated carbon fibers, however, microspores are formed on the surface of thin carbon fibers. Such a direct exposure of microspores to adsorbates gives an advantage of fast adsorption/desorption. The pore development in carbon materials in nano metric scale by air oxidation was studied in detail, which is an activation process with the simplest in the equipments, the mildest in thermal conditions, and also energy and resources saving among different activation processes. The commercially available carbon spheres having the diameter of approximately 15 μm were activated at different temperatures of 355–430 °C for various residence times of 1–100 h in a flow of dry air [146–162].

The original carbon spheres had negligibly small BET surface area, no pores on their surface were observed with high magnification scanning electron microscopy, and their structure was amorphous with non graphitizing random nano texture, so called glasslike carbon. Therefore, oxidation is supposed to start only on the physical surface of each sphere and proceed to the inside of the spheres by forming macrospores, mesopore, and microspores. Each experimental point on oxidation yield against logarithm of residence time, log t, at different temperatures are superimposed on the curve for a reference temperature of 400 °C by the translation along the log t axis to give a smooth curve, being called the master curve. Plot of shift factors against inverse of oxidation temperature can be approximated to be linear and gives apparent activation energy AE of approximately 150 kJ mol^{-1} from its slope. In wet air, however, AE of approximately 200 kJ mol^{-1} was obtained at the same temperature range. For different pore parameters measured by BET, as BJH, and DFT methods, the master curves could be obtained by applying the same shift factors. Master curves for microspore volume determined by as plot, ultra microspore and super microspore volumes by DFT method and mesopore volume by BJH method are shown in order to make the comparison easier, together with some SEM images to show the appearance of the spheres. In the beginning of oxidation, up to 10 h oxidation at 400 °C, the main process is the formation of ultra microspores. Above 10 h up to approximately 60 h, relative amount of ultra microspores decreases but super microspores increase with increasing oxidation time. Above 65 h, both ultra microspores and super microspores decrease rapidly but mesopore increases slightly, which result in the decrease in surface areas. This change in pore volumes might suggest the gradual enlargement of pore size from ultra microspore through super microspore to mesopore and possibly to macrospores by prolonged oxidation above 100 h. At very beginning of activation, 1 h oxidation at 400 °C, BET surface area reaches 400 m²/g of which the predominant part is micro porous surface area. This was experimentally proved to be caused mainly by the change of closed pores, which were formed during carbonization of the precursor phenol resin, to open pores. To understand the activation process from the view point of gasification, it was proposed to normalize the fractional weight loss in different atmospheres (different oxidizing agents, such as steam and CO_2, and their different pressures) as a function of t/t_o 5, where to is the time giving fractional weight loss of 0.5. The experimental data of mass loss obtained at a constant temperature for each oxidizing agent were successfully unified to one curve. The activation process of glasslike carbon spheres in air at different temperatures and residence times was shown to be understood by master curves for the yield and pores structure parameters. On the same carbon spheres, adsorption behaviors of various organics in their aqueous solutions were also understood by the master curves for each adsorbates as functions of oxidation temperature and time. Unification curves are shown as a function of dimensionless time t/t_o 5 and master curves are expressed by real time at a reference temperature. The former seems to be useful to compare the activation (gasification) of various carbonaceous materials and to discuss its mechanism, but the latter might be useful to discuss the activation conditions to prepare activated carbons [147–162].

The derivation procedure of master curves suggests that the conversion between oxidation temperature and time was possible for pore structure parameters as well as activation yield through air oxidation. Activation processes have been pointed out to have some demerits. The mesopore can usually be created by the enlargement of microspores, in other words, by some expense of microspores, and certain part of carbon atoms has to be gasified to CO and/or CO_2 during activation process, in other words, the final yield of activated carbons becomes low. These demerits were pointed out to be one of the barriers for cost down of the industrial production of activated carbons. Also these demerits of activation process were one of the motive forces for the development of new preparation processes of porous carbons, as described in the following sections. Activation in two steps in air was reported to be efficient, the first step at a high temperature for a short time, followed by the second step at a low temperature for a long residence time. On glass like carbon spheres, two-steps activation at 500 °C for 3 h for the first step and at a temperature below 415 °C for different periods for the second step. In the beginning of activation, activation yields larger than 60wt%, higher S_{BET} and S_{mw} are obtained with the same yield, in other words, the same surface areas can be obtained with about 10 wt% higher yields by two-step activation than by one-step activation [162].

1.11.2   TEMPLATE METHODS

Micro porous carbons, of which the highest surface area and pore volume were approximately 4000 g^{-1} and 1.8 mL/g, respectively, were prepared through the carbonization of a carbon precursor in nano channels of zeolites, with the procedure called as template carbonization technique. Since the size and shape of the channels in zeolites are strictly defined by their crystal structures and the pores formed in the resultant carbon are inherited from their channels, microspores formed in the resultant carbons are homogeneous in both size and morphology. Impregnation of fruitfully alcohol (FA) into the channels of zeolites was carried out under vacuum at a low temperature, followed by washing excess FA attached on physical surface of zeolites particles with mesitylene. The composite particles of FA/zeolites thus obtained were heated at 150 °C for 8 h to polymerize FAs (PFA/zeolites composite). The composites were carbonized at 700 °C, followed by dissolution of the template zeolites by 46–48% HF solution. The relation of zeolites cage to the resultant pores in carbon. Detailed preparation procedure for these highly micro porous carbons was reviewed. High-resolution TEM images show that regular alignment of super cages with the size of 1.4 nm is inherited in the resultant carbon as the periodicity of approximately 1.3 nm. The carbon gives a diffraction peak at an angle of approximately 6° in 2θ of CuKa X-rays, just as the template zeolites does, which corresponds to the periodicity of approximately 1.3–1.4 nm due to super structure. Very high surface area indicates the presence of the curved carbon surfaces, which was predicted by grand canonical Monte Carlo simulation. Mesoporous carbons were also prepared by template method using mesoporous silica. By coupling with an activation process, microspores could be easily introduced into these mesoporous carbons [21, 40, 162].

A simple heat treatment of thermoplastic precursors, such as poly (vinyl alcohol), hydroxyl propyl cellulose, polyethylene terephthalate, and pitch in the coexistence of various ceramics, was able to coat ceramic particles by porous carbon. Using MgO particles as substrate ceramics, it was experimentally proved that the carbons formed from PVA at 900 °C were experimentally shown to have larger surface area after the substrate MgO was dissolved out by a diluted acid to isolate carbon. The carbons have high S_{BET}, particularly the carbons obtained from the mixtures prepared from Mg acetate and citrate with PVA in their aqueous solutions (solution mixing). Most of carbons show a relatively high external surface area next which is known to be mostly due to mesopore. Mesopore formed in the carbon was known to have almost the same size as MgO particles, which were formed by the thermal decomposition of Mg compounds in advance of the thermal decomposition and carbonization of carbon precursor. When MgO particles with rectangular morphology were used, the rectangular pores with the same size were observed under SEM in the resultant carbons. Therefore, it was concluded that MgO works as a template for mesopore formation. Pore size distributions in microspore and mesopore ranges measured by DFT and BJH methods, respectively. The carbons formed on the surface of MgO particles are micro porous, rich in the pores with the width of approximately 1 nm. The size of mesopore formed in the carbon is found to depend on the starting MgO precursors, Mg acetate mixed with PVA in solution forming mesopore with the size of approximately 10 nm, Mg citrate those with 5 nm, and Mg gluconate those with 2–4 nm. In the case of Mg acetate/PVA mixtures, the sizes of mesopore depend strongly on mixing method; mixing in powder (powder mixing) gives a broad distribution of pore size but solution mixing gives a relatively sharp distribution of mesopore at around 10 nm. However, Mg citrate/PVA mixtures gave almost the same size distribution of mesopore by mixing in either powder or solution. The substrate MgO had an additional advantage that it could be easily dissolved out at room temperature by a diluted solution of acid, even by 1 mol/L citric acid, so that MgO was experimentally proved to be recycled. The preparation of mesopore carbons using MgO template and their applications were reviewed. Porous carbons containing both microspores and mesopore were also prepared by using Ni hydroxide template. Aqueous suspension of $Ni(OH)_2$ which was prepared from $Ni(NO_3)_2$ and NaOH, was mixed with ethanol solution of phenol resin and then carbonized at 600 °C after drying. By dissolving inorganic species, formed during the process (NiO and Na_2CO_3), porous carbon was isolated. The carbon obtained had S_{BET} of 970 m^2/g, total pore volume of 0.69 mL/g and micros pores volume of 0.3 mL/g in which the predominant microspores and mesopore sizes were approximately 0.8 nm and 15 nm, respectively [26, 162].

1.11.3   DEFLUORINATION OF PTFE

Porous carbons were reported to be prepared through de fluorination of poly tetra fluoro ethylene (PTFE) with alkali amalgamates. The detailed studies were reported for the preparation of porous carbons from PTFE by using different alkali metals. PTFE film was pressed with lithium metal foil under 4 MPa in Ar atmosphere for 48h to de fluorinate PTFE. After excess lithium metals were washed out with methanol, the heat treatment at 700 °C and washing by dilute HC1 were carried out in order to eliminate finely dispersed Li F. De fluorination of PTFE was also possible through heating a mixture of PTFE powders with alkali metals, Na, K, and Rb in vacuum at 200 °C in a closed vessel. N_2 adsorption isotherms of resultant carbons were found to depend strongly on alkali metal used. De fluorination of PTFE with Na metal was found to give mesopore rich carbon and very high S_{BET} as 2225 m^2/g. S_{BET} of these carbons prepared using Na was found to increase with heat treatment at a high temperature up to 1000 °C, probably because of gasification of carbon with surface oxygen functional groups. De fluorination of PTFE using Na metal has an advantage, Na metal is much cheaper in price and easier to handle, and so the process being simpler than using other alkali metals.

The irradiation of PTFE before carbonization is also preferable for getting high surface area. De fluorination of PTFE was possible in 1,2-dimethoxyethane (DME) solution of alkali metals naphthalene complexes at room temperature. S_{BET} of the resultant carbon was 1000–1800 m^2/g and no effect of alkali metals on pore structure was observed [162].

1.11.4 CARBON AEROGELS

Carbon aero gels, which have been well known as one of mesopore carbons, were prepared from the pyrolysis of organic aerogels of resorcinol and formaldehyde. Extensive studies focused on their pore structure and also on doping of some metals were carried out. Primary carbon particles have the size of approximately 4–9 nm and interconnected with each other to forma network. Adsorption isotherms belong to type IV and have a clear hysteresis. Pore structure parameters calculated through analysis are listed up on carbon aero gels prepared from resorcinol and formaldehyde at different temperatures can report. These carbons aero gels contain predominantly inter particle mesopore formed in a three-dimensional network of interconnected minute carbon particles, and only small amount of intra particle microspores were formed in primary carbon particles. Carbon aerogels could be activated in order to increase microspores by CO_2 at 900 °C. The activation for 5 h increased both of microspores and mesopore: pore volume of 0. 68 and 2. 04 mL g and surface area of 1750 and 510 m^2/g, respectively. The detailed studies on adsorption of N_2 at 77 K and of water vapor at 303 K on activated carbon aerogels, whose surface functional groups were showed clearly that the amount of adsorbed water corresponded only to the microspore volume and not to the mesopore volume. Doping of Ce and Zr into carbon aero gels was found to result in micros pore rich carbon materials. Pore structure of carbon aero gels was known to be governed by that of precursor organic aero gels, which was controlled by the mole ratios of resorcinol to formaldehyde (R/F), to water (R/W) and also to basic catalyst Na_2CO_3(R/C). Aqueous gels synthesized were dried under supercritical CO_2. Pore size distributions in mesopore region for both original organic aero gels and resultant carbon aero gels as a function of R/W, the other factors R/F and R/C being the constant as 0.5 and 75, are reported. Pore size distributions of organic aero gels are rather sharp and their maxima decrease with increasing R/W ratio. By carbonization of these organic aero gels, pore size distributions shift to a little smaller size, mainly because of the shrinkage of gels during thermal decomposition. Instead of supercritical drying of aqueous gels, freeze drying method was also applied. On the gels prepared through freeze drying (cryogels), much smaller shrinkage in pore size during carbonization was observed. Conditions for the preparation of resorcinol-formaldehyde gels through sol-gel condensation and those for freeze-drying were studied in detail in order to control mesoporosity in resultant carbons. Effect of drying process of gels, such as freeze drying, microwave drying, and hot air drying, on pore structure were also studied. The first two drying methods are effective in order to get mesoporous carbons. Resorcinol-acetaldehyde cryogels could be also a precursor for mesoporous carbons. Pore structure in the resultant carbons was found to depend on pH-value in the solution, which was changed by mixing ratio of R/C. Pore volumes of the resultant carbon aerogels after 800 °C carbonization with pH of 8.0 (R/C = 25), no porous carbon is obtained. Porous carbons are formed in the pH range of 7.0–8.0, particularly mesopore-rich in the pH range of 7.3–7.7. Carbon aerogels were used as a template for the preparation of highly crystalline zeolite with uniform mesopore channels [1, 2, 39, 162].

1.11.5 POLYMER BLEND METHOD

Polymer blend method was proposed to synthesize various types of carbons, mixing two different polymers, one having a high carbon yield, such as polyfurufuryl alcohol, and the other, a low carbon yield, such as polyethylene. The scheme of polymer blend method to get carbon balloons, carbon beads, and also porous carbons is shown in Fig. 1.12. By applying spinning on blended polymer, certain success in obtaining carbon nano fibers was reported. Through the synthesis of polyurethane-imide films and their carbonization, carbon films were obtained, of which macro pore structure was controlled by changing the molecular structure of polyurethane. Pre-polymer poly (urethane-imide) films were prepared by blending polyamide acid giving polyimide (PI) with polyurethane (PU) Polyurethane-imide films after heating up to 200 °C showed phase separation of PI and PU, where the former polymer formed a matrix and the latter formed small islands. By heat treatment up to 400 °C, PU component was pyrolyzed to gases and resulted in porous PI films, which can be converted easily to porous carbon films by carbonization. Pore sizes in these carbon films were controlled by the blending ratio of PI to PU and also by the molecular structure of PU [1, 2, 162].

1.11.6 SELECTION OF SPECIFIC PRECURSORS

1.11.6.1 DERIVATIVES OF BUTADIYNYLENE

Poly (phenylene butadiynylene) were found to give very high carbon yield of more than 90wt% after the heat treatment at 900 °C, very close to theoretical yield, whose molecular structure are determined. The resultant carbons are amorphous state and micro porous, having total surface area of 1–330 m^2/g micro porous surface area of 1–300 m^2/g and microspore volume of 0.49 mL/g. Their pyrolysis behavior was characterized by a very sharp exothermic peak at around 200 °C without accompanying any mass change, which was due to 1,4-polymerization of buta diynylene moiety and resulted in cross-linking between molecules, and a little but gradual mass loss of approximately 600 °C. The material heat-treated above 200 °C was so highly and strongly cross linked that the hydrogen atoms that remained were mostly stripped off as hydrogen molecules, which is the main reason to give high carbon yield. The derivatives with methyl radicals on benzene ring gave also high carbon yields, close to 80wt% of carbon content [1, 162].

1.11.6.2 FLUORINATED POLYIMIDES

Micro porous carbon films were prepared from aromatic polyimides synthesized from different dianhydrides and diamines with different molecular structures by carbonization. Structures of repeating unit in polyimides are reported. They were characterized by the presence of pending groups, –CH, and –CF, and number of phenyl ring in the repeating unit. Highly micro porous carbon films were obtained without any activation process, of which micro porous surface area was closely related to the contents of phenyl ring [1, 2].

1.11.6.3 METAL CARBIDES

Various metal carbides were found to give highly micro porous carbon through the heat treatment in a flow of carbide-derived carbons. S_{BET} was reported to be 1000–2000 m^2/g^{-1}, depending on the precursor carbide and heat treatment temperature. The carbide-derived carbons gave a relatively sharp distribution in pore size and their predominant pore size depends mostly on the precursor carbide; SiC, TiC and Al_4C_3 gave 0.7–0.8 nm, and B_4C approximately 1.3 nm and Ti_3SiC_2 0.5–0.6 nm. From ZrC_{098}, however, the carbons were obtained with a sharp size distribution of approximately 0.7 nm after the heat treatment at 300–600 °C, but with a broad distribution ranging from 0.6 to 3 nm after the heat treatment at 800–1200 °C. These carbide-derived micro porous carbons were used as the electrode of electric double layer capacitors [162].

1.11.6.4 BIOMASSES

The cores of kenaf plant (Hibiscus cannabinus) were found to give a high S_{BET} as high as 2700 m^2/g by the heat treatment, without any additional activation process. This high surface area was supposed to be due to the departure of metallic impurities (mostly K), which were originally included in the cores, during carbonization. Porous carbons were successfully prepared from chips of cypress (Cupressus) by the heat treatment under a flow of superheated steam. Pore structure in cypress charcoals could be controlled by changing carbonization conditions (temperature of superheated steam, and supplying and transferring rates of chips) [162].

1.11.7 SELECTION OF PREPARATION CONDITIONS

1.11.7.1 MOLECULAR SIEVING CARBON FILMS

Carbon membranes with molecular sieving performance were successfully prepared by the carbonization of commercially available polyimide films with the thickness of 0.1 mm. adsorbed amounts of gas molecules with different sizes are shown on the carbon films heat treated at different temperatures, ethane C_2H_6 with the size of 0.40 nm permeating in a large amount through the film heated up to 700 °C, but with very small amount through the one heated up to 1000 °C. This result suggested a strong dependence of pore sizes in carbon films on carbonization temperature, which was supposed to be due to shrinkage during carbonization [1, 2, 54, 55].

1.11.7.2   MACRO POROUS CARBONS (CARBON FOAMS)

Carbon foams were prepared from mesophase pitch through either blowing or pressure release of molten pitches followed by stabilization in air. Because the foam with a high thermal conductivity attracted attention for a potential filler material of some composites, the preparation of a large sized carbon foam with high thermal conductivity from mesophase pitch by a new and less time consuming process was extensively studied. The graphitized foams have the bulk density of 0.2–0.6 g cm^{-3} with an average pore size of either 275–350 micrometers or 60–90 micrometers and relatively high graphitization degree [162].

1.11.7.3   INTERCALATION

Intercalation of various species into graphite gallery can be considered to be the chemical adsorption into intrinsic pores, which are flexible and two-dimensional slit-shaped microspores. These intercalation phenomena are one of characteristics of carbon materials in graphite family. In addition, intercalation can be understood as a new process to create extrinsic pores in the compounds, of which the size is controllable by the size of intercalates. In graphite intercalation compounds (GICs) of alkali metals with 2-D structure, for example, nano spaces are formed surrounded by alkali metal ions and graphite layer planes. As an example, arrangement of Cs^+ ions in graphite gallery. In this compound, the space with the size of approximately 0.311×0.266 nm can accept the third component. Changing the alkali metal ions to K (ion radius of 0.133 nm) and Rb (0.148 nm) from Cs^+ (0.169 nm), the size of the space for the third component changes [1, 2, 162].

1.11.7.4   EXFOLIATION

Exfoliated graphite has been produced as a raw material for flexible graphite sheets in a huge amount in the world. Usually it is produced through rapid heating of residue compounds of natural graphite flakes with sulfuric acid up to a high temperature as 1000 °C. It consists of fragile worm like particles, which are formed by exfoliating preferentially along the normal to the basal plane of graphite. There are at least three kinds of pores in exfoliated graphite. Large pores are formed mainly by the complicated entanglement of wormlike particles during abrupt exfoliation of each graphite flakes and so they are flexible inter particle macro pores. In addition, crevice like pores on the surface of worm-like particles and pores inside the particles are formed respectively. In order to evaluate the pores, which are rigid intra particle pores, a technique to prepare fractured cross-section of worm like particle had to be developed and various pore structure parameters were determined with the aid of image processing technique [162].

1.12 IMPORTANCE AND NECESSITY OF CONTROLLING OF PORE STRUCTURE IN CARBON MATERIALS

1.12.1 IRREVERSIBLE ADSORPTION OF CO_2 GAS

A glass-like carbon sphere was found to show an irreversible adsorption of CO_2 gas. Two commercially available glass-like carbon spheres APS and APT were used, which were prepared at 700 °C and 1000 °C, respectively, in CO_2 atmosphere. The size of the spheres was rather homogeneous, at approximately 15 micrometers. The adsorption/desorption behavior of CO_2 was measured volumetrically at different temperatures and gravimetrically at 0 °C. When CO_2 pressure in the electro balance was less than 1atm, the adsorbed amount of CO_2 at the saturation on APT decreased, although the behavior during adsorption, evacuation, and heating was the same. Selective adsorption of CO_2 was shown to be possible by the carbon spheres after the irradiation of oxygen plasma. These gravimetric studies on the adsorption/desorption of CO_2 suggest the presence of three types of microspores in APT, depending on the accessibility of CO_2. The first type of microspores adsorbs CO_2 molecules immediately and also desorbs quickly only by evacuation at 0 °C, of which the volume is supposed to be approximately 40%. The second type of microspores adsorbs CO_2 very slowly, taking approximately 200h to reach saturation and requires heating to 250 °C under vacuum for desorption, being about 55%. From the third type of microspores, the adsorbed CO_2 is not released even by heating up to 250 °C under vacuum (strongly trapped CO_2 about 5% in volume). The detailed study by supercritical gas adsorption analysis with the aid of grand canonical Monte Carlo simulation showed that the width of entrance (mouth) of ultra microspores in APT is a little narrower than that of inside of pores [1, 2, 162].

1.12.2   CARBON FOAM FOR WATER VAPOR ADSORPTION/ DESORPTION

In order for micro porous carbons to play certain functionality, microspores are preferred to be directly open to adsorbents in most applications for their easy and rapid adsorption/desorption. To keep a large number of microspores open, morphology of carbon materials is important, as mentioned on activated carbon fibers. Carbon foams can also satisfy this requirement. Carbon foams were prepared from a fluorinated polyimide (6FDA/TFMB), which was known to give micro porous carbon, by using either urethane or melamine foam as a template and by subsequent activation in air at 400 °C. Their adsorption/desorption behaviors for water vapor were examined in TG apparatus. Many points are remained to be studied; optimization of pore structure for adsorption/desorption of water vapor with higher rate and also at lower relative pressure, search for more appropriate template foam and impregnate polymer, etc. [162].

1.12.3   POROUS CARBON FOR CAR CANISTER

A strong demand to recover gasoline vapor during parking of cars is now understood to be important for saving gasoline and also for avoiding the contamination of air. Gasoline vapor is absorbed into activated carbon in a device called "canister" during parking and reused by passing air through adsorbed activated carbon during running. For activated carbon in the car canister, a specific pore structure is required. For the evaluation of the performance of car can aster, adsorption/desorption of butane gas has often been used. In the recent work [1, 162], direct gravimetric measurement of adsorption of gasoline vapor was carried out, which was not a standardized method for car canister, and also no experiment on desorption of gasoline vapor was done. The confirmation of the results is demanded by using butane gas.

1.12.4 CARBON ELECTRODE FOR ELECTRIC DOUBLE LAYER CAPACITORS (EDLC)

As mention in Section 1.1.4, because EDLC is on the basis of the formation of electric double layer at the electrode/electrolyte interface, which is due to the physical adsorption of electrolyte ions, activated carbons with high surface area are usually used as electrode materials and their PSD is pointed out. To have an influence in EDLC performance Capacitance of EDLCs was possible to be explained by dividing into two capacitances of the surface due to microspores (micro porous surface area) and of that due to larger pores (evaluated as external surface area) on different activated carbons. Porous carbons are characterized by extremely large BET surface areas that range from 500 to 3000 $m^2 g^{-1}$. This surface area largely arises from a complex inter connected network of internal pores. Microspores have a high surface area to volume ratio and, consequently, when present in significant proportions is a major contributor to the measured area of high surface area activated carbons. Microspore sizes extend down to molecular dimensions and play an important role in the selectivity of adsorption-based processes, through restricted diffusion and molecular sieve effects. Fine microspores also exhibit a greater adsorbent adsorb ate affinity due to the overlap of adsorption forces from opposing pore walls. Accordingly, adsorption in fine pores can occur via a pour filling mechanism rather than solely by surface coverage (as is assumed by the Langmuir and BET calculations of surface area). In such cases, the conversion of adsorption data into an estimate of surface area, by the application of the BET equation, can lead to unrealistically high surface area estimates. Clearly, the pore size distribution of porous carbons influences to a large degree the fundamental performance criteria of carbon-based super capacitors, the relationship between power and energy density, and the dependence of performance on frequency [20, 34, 45, 162].

Not surprisingly, therefore, considerable research is presently being directed towards the development of carbon materials with a tailored pore size distribution to yield high capacitance and low resistance. Electrodes EDLCs are expected to be one of the promising devices for electric energy storage and a variety of applications have been developed, for example, for memory back-up, cold start vehicle assist, storage for solar cell power, and also for high power sources, such as power trains and electric vehicles. Extensive research works are still carried out in order to obtain better performance. Asymmetric EDLCs using different activated carbons in two electrodes, negative and positive electrodes, were constructed and their performance was studied in nonaqueous electrolyte. The capacitance and rate performance of these asymmetric EDLCs were found to be governed predominantly by the pore structure of the carbon in the negative electrode [108, 162].

1.12.4.1 SUPER CAPACITORS BASED ON ACTIVATED CARBON MATERIAL

More than 20 years ago an experimental super capacitor cell by using commercial ACF cloth for each of the two electrodes and glass fiber filter paper as separator in organic electrolyte was realized. At that time 36.2Wh/kg specific energy, 11.1 kW/kg specific power and 36.5 F/g specific capacitance were estimated. The specific capacitance was considered per gram of ACF. For the ACF cloth made from phenol resin, a specific surface area of 1500–2500 m^2/g was estimated. The specific energy and specific power reached in practice at this time are much lower than the above estimated values because these take into consideration the overall weight of capacitor cell including its package. Activated carbon composite electrodes for electrochemical capacitors have been also investigated. Thus for hydrous ruthenium oxide/activated carbon electrode in H$_2$SO$_4$ electrolyte, an increase of specific capacitance from 243 F/g (for pure activated carbon electrode) to 350 F/g for composite electrode where 35% is ruthenium oxide is reported [162].

In other work for only 3.2% ruthenium oxide in the composite electrode, an increase in the capacitance of 25% to a value of 324F/g is reported. Other experiments with ruthenium oxide/activated carbon composites used at positive electrodes in electrochemical capacitors indicated increase of the specific capacitance. Nickel hydroxide/activated carbon composite electrodes used in electrochemical capacitors provide significant increase of specific capacitance from 255 to 314F/g. If manganese oxide/activated carbon composite electrodes are used increase of specific capacitance takes place. Other materials used for activated carbon composite electrodes have been found to increase the specific capacitance. In hybrid or asymmetric electrochemical capacitors, one electrode is based on activated carbon material and another one is based on another material (nickel hydroxide, manganese oxide, etc.). Higher specific capacitance or specific energy is possible than in the case of symmetric capacitors, based only on activated carbon electrodes [34, 45, 64].

It is usually anticipated that the capacitance of a porous carbon (expressed in F g^{-1}) will be proportional to its available surface area (in m^2 g^{-1}). Whilst this relationship is sometimes observed, in practice it usually represents an oversimplification. The major factors that contribute to what is often a complex (nonlinear) relationship are: (i) assumptions in the measurement of electrode surface-area; (ii) variations in the specific capacitance of carbons with differing morphology; (iii) variations in surface chemistry (wet ability and pseudo capacitive contributions); and (iv) variations in the conditions under which carbon capacitance is measured. The surface areas of porous carbons and electrodes are most commonly measured by gas adsorption (usually nitrogen at 77 K) and use BET theory to convert adsorption data into an estimate of apparent surface area. Despite its widespread use, the application of this approach to highly porous (particularly micro porous) and heterogeneous materials has some limitations and is perhaps more appropriately used as a semiquantitative tool. Possibly the greatest constraint in attempting to correlate capacitance with BET surface area, is the assumption that the surface area accessed by nitrogen gas is similar to the surface accessed by the electrolyte during the measurement of capacitance. While gas adsorption can be expected to penetrate the majority of open pores down to a size that approaches the molecular size of adsorb ate, electrolyte accessibility will be more sensitive to variations in carbon structure and surface properties. Electrolyte penetration into fine pores, particularly by larger organic electrolytes, is expected to be more restricted (due to ion sieving effects) and vary considerably with the electrolyte used. Variations in electrolyte–electrode surface interactions that arise from differing electrolyte properties (viscosity, dielectric constant, dipole moment) will also influence wet ability and, hence, electrolyte penetration into pores [64–68].

1.12.4.2 SUPER CAPACITORS BASED ON CNTS

The presence of mesopore in electrodes based on CNTs, due to the central canal and entanglement enables easy access of ions from electrolyte. For electrodes built from MWCNTs, specific capacitance in a range of 4–135

F/g was found in Refs. For SWCNTs, a maximum specific capacitance of 180F/g and a measured power density of 20kW/kg at the energy density of 7Wh/kg, in KOH electrolyte is reported. In other work an initial specific capacitance of 128 F/g decreased after charging–discharging cycles to 58 F/g. Enhancement of specific capacitance given by CNTs is possible for example by their mixing with conducting polypyrrole. A comparative investigation of the specific capacitance achieved with CNTs and activated carbon material reveals the fact activated carbon material exhibited significantly higher capacitance. SWCNTs/polypyrrole nanocomposite electrode used in recently work indicated much higher specific capacitance than pure polypyrrole or SWCNTs electrodes. Super capacitor electrodes based on CNT–PANI nanocomposite by coating poly aniline on the surface of the CNT have been used recently. At a current density of 10 mA/cm, the CNT–PANI nanocomposite exhibits high specific capacitance of 201 F/g, in comparison with a value of 52 F/g for the CNT. The super capacitors based on the CNT–PANI nanocomposite have an energy density of 6.97 Wh/kg and an outstanding power performance [65, 162].

1.12.5   MACRO POROUS CARBON FOR HEAVY OIL SORPTION AND RECOVERY

Exfoliated graphite was found to sorbs a large amount of heavy oils at room temperature very quickly, more than 80 kg^{-1} within 1 min. Different carbon materials were studied through the determination of sorption capacity, sorption kinetics, and repeated sorption/desorption cycles in order to recover spilled heavy oils and also to recycle both heavy oils and carbon materials. Exfoliated graphite was then applied to sorbs other oils, such as cooking and engine oil sand also to organics with large molecules, such as bio-fluids. Exfoliated graphite with a low bulk density has very high sorption capacity for heavy oil, but its sorption rate is rather low. By increasing its bulk density, the sorption rate can be improved slightly, but sorption capacity decreases quickly at the same time. Carbonized fir fibers have a sorption capacity comparable with the exfoliated graphite and the same dependence on bulk density. By densification of carbonized fiber felts, however, sorption rate increases rather rapidly. These high sorption capacities for the carbon materials were found to be mainly because of flexible macro pores in these carbon materials. Macro pores with the size in the range of 1–600 micrometers are primarily responsible for heavy oil sorption [1, 162].

1.13 RECENTLY STUDY WORKS ABOUT CONTROLLING OF PORE SIZE

Summary of some recently reported papers about confirmed modeling of controlling of pore size in carbon-based nano adsorbent are presented in Table 1.4 [8, 40, 121, 161].

TABLE 1.4 Summary of Recently Confirmed Models for Controlling of Pore Size in Carbon Based Nano Adsorbents

Carbon Material Type	Applied Model and Simulation Methods	References
AC, ACF, MSC	N_2 adsorption at 77 K and t-plot, Alpha-plot	[8]
Carbon structures	N_2 adsorption at 77 K and Alpha-plot	[9]
AC	G CMC and DFT (Monte Carlo simulations)	[10]
Carbon structures	N_2 and NLDFT model	[11]
AC	G CMC	[12]
ACF	H_2S adsorption	[13]
C-ZIFs	N_2 adsorption at 77 K and t-plot, BET	[14]
Carbon structures	ND and DFT	[15]
AC	N_2 adsorption at 77 K and t-plot, Alpha-plot	[16]
AC	Monte Carlo Simulations	[17]
p-carbon	NLDFT-BJH method	[18]
ACF and AAPFs	DR equation	[19]
AAC	N_2 adsorption at 77 K and t-plot, Alpha-plot	[20]
ACH structure	N_2 adsorption at 77 K and t-plot, Alpha-plot	[21]
ACF	N_2 adsorption at 77 K and t-plot, Alpha-plot	[22]
AC	Ar adsorption at 87 K and t-plot	[23]

TABLE 1.4   *(Continued)*

Carbon Material Type	Applied Model and Simulation Methods	References
ACF	Ar adsorption at 87 K and Alpha-plot	[24]
AC	IAST-Freundlich model	[25]
ACF	Ar adsorption at 87 K and t-plot, BET	[26]
MSC, AC	N_2 adsorption at 77 K and t-plot, Alpha-plot	[27]
P-ACS	N_2 adsorption at 77 K and t-plot, BET	[28]
AC	GCMS and SIE equation – methane adsorption	[29]
AC	N_2 adsorption at 77 K and t-plot, Alpha-plot	[30]
Carbon structures	Ar adsorption at 87 K and t-plot, Alpha-plot	[31]
Carbon structures	N_2 adsorption at 77 K and t-plot, Alpha-plot	[32]
Carbide-derived carbons	Ar adsorption at 87 K and t-plot	[33]
C-xerogels	N_2 adsorption at 77 K and Alpha-plot, BET	[34]
Carbide-derived carbons CDC's	N_2 adsorption at 77 K and t-plot, Alpha-plot	[35]
Carbon black	NLDFT	[36]
Carbon Structures	N_2 adsorption at 77 K and t-plot, Alpha-plot	[37]
Glassy carbon	Monte Carlo Simulation N adsorption	[38]
Carbon structures	DFT–Ar adsorption at 77 K	[39]
AC	N_2 adsorption and DR equation–BjH	[40]
P-ASC	N_2 adsorption at 77 K – BjH method	[123]
ACF	DRS equation-N_2 adsorption at 150 °C	[126]
AC	DFT–N_2 adsorption at 77 K	[128]
CNT/Polymer Composite	MDS–NPT	[132]
CNT composite	MDS –PMF	[121]
SWCNT	GCMC-LJ potential	[131]
CNT	MDS – PEOE algorithm PME	[124]
CNT/PE composite	MDS – Brenner – Newtonian equation	[125]
SWCNT	MDS – DFTB-CPMD-REBO	[127]
CNT	MDS	[122]

TABLE 1.4   *(Continued)*

Carbon Material Type	Applied Model and Simulation Methods	References
CNT/Polymer	MDS – SHAKE algorithm – DL – Poly	[130]
SWCNT	MDS – CPMD-DFT	[129]
AC	DS-HK-IHK	[133]
2D Graphene	ADS – DFT – APMD – PAW	[134]
SWCNT	MDS – Berenner	[135]
CNT ropes	GCMC	[136]
CNT / Sodium	MDS	[137]
CNT	Abintio QMS-SFC	[138]
CNT	MDS – USHER algorithm – LJ Potential	[139]
Nanotube	MDS – PES – Verlet algorithm	[140]
CNF	MDS – GRASP – RFF	[141]
SWCNT	MDS – DFT – B3LYP	[142]
CNT	MDS – LJ potential	[143]
Metal Membranes	EMD – GCMC – LJ potential	[144]
CNT	MDS-LJ–TIP5P	[145]
ACNF	BET	[146]
AC	GCMC–NLDFT-BjH, HK	[147]
Carbon Structures	NLDFT	[148]
Carbon Adsorbents	SHN1, SHN2 algorithm	[149]
AC	DA, ND, alpha-plot, GCMC	[150]
Carbon microspores	CONTEN, DFT, DA	[151]
AC	ND, HK, DA	[152]
AC	NLDFT	[153]
AC	NLDFT	[154]
Micro and mesopore Carbon	NLDFT, QSDFT	[155]

TABLE 1.4   *(Continued)*

Carbon Material Type	Applied Model and Simulation Methods	References
Micro and mesopore Carbon	GCMC, LJ potential	[156]
AC	ASA algorithm DFT, ND	[157]
Micro and mesopore Carbon	L-Curve method, SHN algorithm	[158]
Micro Carbons	Fulleren Like models for adsorption	[159]
CB – AC	ASA, REG algorithm	[160]
AC	GCMC, LJ potential	[161]

The recently research works about controlling of pore structure in carbon materials that categorized as fallowing include: First, using the grand canonical Monte Carlo (GCMC) method, or other simulation methods to determine adsorption isotherms in Ar (87 K) or N_2(77 K) that are simulated for all the carbon sample structures to reach optimum condition in experimental works. Second experimental works to obtained PSD by t-plot or alpha plot curves and BET equation that determine other adsorption parameters and show that samples structures maybe micro or mesoporous (with different ratio of micro/mesopore). Finally PSD are calculated using the HK, density functional theory (DFT), D-R method, BJH approaches and other mathematical methods and this model results with those predicted by the experimental work results compared and adapted, to prove selected mathematical model is significant and simulation that applied is verified [121, 161].

This review is concerned with such pore controlling methods and models performed by researchers. An ultimate aim of these researchers is to establish a model, which can tailor pore structure of carbon materials to read any kind of requirement. In this review, we would like to highlight how effort the researchers have made to control micro and mesopore in carbon materials, and how active they are in achieving the final target and prepare them for special application that here our means adsorption contaminates from aqueous environments. This is one of the first studies in which different methods of calculation of PSDs for carbon nano structures from adsorption data can be really verified, since absolute PSDs are obtained using the certain method. This is also one of the first studies reporting the results of computer simulations of adsorption on carbon structure models. There are well documented reports in the area of PSD estimation and pore structure control modeling from adsorption data and compare of this results with theoretical predictions; for example, effect of chemical ratio and compare of some type of activating agent on pore structure of carbon materials, with some different classical models, but it appears there is still a lack of studies on the modeling and simulation methods of controlling pore structures in carbon based nano adsorbents, that focus on parameters such as range of applied temperatures in activation state and applied special catalysts in activation condition that affect controlling of pore structure of carbon materials. This field of research has a great room for improvement in pore structure control modeling and simulation methods of carbon materials. There is no definite answer to this argument since each of these adsorbents has its own advantages and disadvantages in their special applications.

KEYWORDS

- **Basic concepts**
- **Carbon nanotubes**
- **Classical models**
- **Micro or mesoporous**
- **Pore controlling methods**
- **Simulation methods**

CHAPTER 2

MATHEMATICAL MODELING

CONTENTS

2.1 INTRODUCTION

Finding a reliable, accurate, and flexible method for the determination of PSD of porous adsorbents still remains an important concern in the area of characterization of porous materials. Although a large number of researches have been done in this area, some constraints such as type of adsorbate, adsorbent characteristics, adsorption temperature, applicable range of pore size, and range of relative pressure limit the applicability of each model in all cases. The lack of such method is tangible by rapid development of new porous materials and their wide applications in various fields. In this chapter, the following three well-known models were used in order to obtain PSD for two series of chemically activated carbons and the results are compared. It is increasingly common to study adsorption processes, whether on free surfaces or in confirmed spaces such as pores by modeling or simulation techniques. The reach aim of these studies is frequently to develop an understanding that will better enable adsorption measurements to be used to characterize various adsorbents in terms of their surface properties or pores structure. Modeling and simulation methods include: Grand Canonical Monte Carlo (GCMC), density functional theory (DFT), LJ potential, BJH, Novel Algorithms method such ASA, Varlet, SHN, HK and IHK method, DR method, DS method (Stoeckeli method), etc. [121, 161].

There is also a different approach based on a single adsorption isotherm. Here, total adsorption amount, which is simply a summation of the adsorbed molecules on various adsorption sites is equal to an integral of the local adsorption on particular sites multiplied by a PSD function, integrated over all sizes:

$$\theta(P) = \int_{0}^{\infty} \theta(L, P) f(L) dL, \tag{1}$$

where θ (L, P) is the local adsorption isotherm (kernel) evaluated at bulk pressure P and local pore size (L), and F(L) denotes the PSD of the heterogeneous solid adsorbent. Solving for the PSD function is an ill-posed problem unless the form of function is defined. Various models by assuming different kernels (Langmuir, Freund lich, BET, DR, DA, Sips, Toth, Unilan, Jovanovich, Fowler and Harkins) and mathematical functions (Gaussian, Gamma) for PSD have been presented. For instance, the

Dubinin-Stoeckli (DS) and Stoeckli models which have been proposed based on the Dubinin theory of volume filling of microspores (TVFM) implement Gaussian and gamma type of mathematical function, respectively [1, 2, 133, 155] (Table 2.1).

TABLE 2.1 Techniques for Characterization of Pores in Carbon Materials

Characterization technique	Comments (advantages and disadvantages)
Adsorption/desorption of N_2 gas at 77 K	
BET method	Give overall surface area (SS).
Alphaplot	Give micro porous and external SSs separately.
	Give microspore volume.
BJH method	Differentiate micro porous and mesopores SSs and volumes.
	Give pore size distribution in mesopore range.
DFT method	Give pore size distribution in a wide range of size.
HK method	Give pore size distribution
etc.	
Adsorption/desorption isotherm of various gases (H_2, He, CO_2, CO, etc.)	Give the information of molecular sieving performance.
X-ray small-angle scattering	Detect microspores, both open and closed pores.
Transmission electron microscopy	Detect nano-sized pores, even less than 0.4 nm size.
	Give localized of information, need statistical analysis of data.
Scanning tunneling microscopy	Detect only pore entrances on the surface.
	Give morphological information of the pore entrance.
	Need statistical analysis with criteria.
Scanning electron microscopy	Detect only macrospores.
Mercury porosimetry	Detect mostly macrospores.
	Difficult to apply for fragile materials.

2.2   BASIC MODELS FOR ADSORPTION PARAMETERS CALCULATING IN CARBON NANO ADSORBENTS

2.2.1   CALCULATE OF S_{BET}, V_{MICRO} AND S_{MESO} IN POROUS CARBONS

First stage in characterization of adsorption properties of active carbons is usually determination of their surface area and pore volume. The surface area is normally determined from equilibrium adsorption isotherm of a gas or vapor measured in a range of relative pressures from 0.01 to 0.3. Currently, there are two major methods used to evaluate specific surface area from gas adsorption data: the Brunauer-Emmett-Teller (BET) method and the comparative plot analysis.

The evaluation of the specific area by the BET method is based on the determination of the monolayer capacity (i.e., the number of adsorbed molecules in the monolayer on the surface of a material) by fitting experimental gas adsorption data to the BET equation:

$$a = \frac{a_m C \frac{p}{p_0}}{\left(1 - \frac{p}{p_0}\right)\left[1 + (C-1)\frac{p}{p_0}\right]} \qquad (2)$$

where is the total amount adsorbed, is the monolayer capacity, p/p_0 is the relative pressure and is the constant related to the heat of first-layer adsorption:

$$C = \exp\left(\frac{q_1 - q_L}{RT}\right) \qquad (3)$$

where is the difference between the heat of adsorption in the first layer and heat ofcondensation, T is the absolute temperature and R is the universal gas constant. The Eq. (3) was derived for an infinite number of adsorbed layers. This equation is usually expressed in the following linear form:

$$\frac{\frac{p}{p_0}}{a\left(1 - \frac{p}{p_0}\right)} = \frac{1}{a_m C} + \frac{C-1}{a_m C}\frac{p}{p_0} \qquad (4)$$

The Eq. (4) is used to evaluation the monolayer capacity, am, which is necessary for the evaluation of the surface area,. If the cross-sectional area for a single molecule in the monolayerformed on a given surface is known, the surface area can be evaluated by using the following formulae:

$$S_{BET} = a_m \omega N_A \qquad (5)$$

where N_A is the Avogadro number. The derivation of the BET equation involves the following major assumptions: the surface is flat, all adsorption sites exhibit the same adsorption energy; there are no lateral interactions between adsorbed molecules; the adsorption energy for all molecules except those in the first layer is equal to the liquefaction energy; and an infinite number of layers can be formed. In the case of adsorption on active carbons, some of these assumptions are often not valid. In particular, surfaces are geometrically and energetically heterogeneous, there are lateral interactions between adsorbed molecules, and interactions of adsorbed molecules vary with the distance from the surface. Therefore, one should not expect the monolayer capacity evaluated by the BET method to be particularly accurate. In addition, the available values of cross sectional area, even in the case of the most commonly used adsorbates, are somewhatuncertain and may actually vary from one type of the surface to another [1, 2, 162].

Moreover, the determination of the specific surface area based on the molecular size and monolayer capacity should be treated with some caution, because the ability of molecules to effectively cover the surface depends on the molecular size and surface roughness. Since adsorbed molecules cannot satisfactory probe the surface on the scale smaller than their size, the surface area determined using larger molecules might be smaller than that obtained from adsorption data for smaller molecules. Despite all these problems and limitations, the BET method is currently a standard way for evaluation of the specific surface area of solids. For several reasons, nitrogen (at 77 K) is generally considered to be the most suitable adsorbate for surface area evaluation and it is usually assumed that the nitrogen monolayer is close packed. For many years nitrogen adsorption data at 77 K have been used to characterize the porous structures of a variety of materials and reference nitrogen adsorption isotherms for different nanoporous carbons have been reported. However, it is possible to characterized nanoporous carbons by using other adsorbents, some of which may be more convenient to use and provide a better insight into the porous structures. Consequently, reference adsorption data on carbons have been published for argon and n-butane, benzene, neo pentane and methanol.

In particular, argon deserves much attention, because it was found to be convenient for the characterization of micro porous and mesopores. The intercept on the adsorption axis evaluated by the back extrapolation of the -plot provides the micro pore adsorption capacity:

$$a\left(\frac{p}{p_0}\right) = a_{mi}^0 + \eta\alpha_s$$
(6)

The micro pore volume V_{mi} cans be evaluated according to the formula:

$$V_{mi} = a_{mi}^0 F$$
(7)

In latter equation, F is a conversion factor (F = 0.0015468 for nitrogen at 77 K, F = 0.001279 for argon at 77 K and 87 K, when the amount adsorbed is expressed in $cm^3\ g^{-1}$ and the pore volume is expressed in $cm^3\ g^{-1}$). The slope of the linear segment of the -plot () permits the evaluation of the monolayer adsorption capacity for the mesopore surface:

$$a_{me}^0 = \eta\frac{a_s^0}{a_s(0.4)}$$
(8)

where is the monolayer capacity of the reference adsorbent, as (0.4) is the amount adsorbed on the nonporous reference adsorbent at relative pressure p/p_o = 0.4. The mesopore surface area is obtained by multiplying by Avogadro's number N_A and the molecular area occupied by one molecule adsorbed on the mesopore surface, that is:

$$S_{me} = a_{me}^0 N_A \omega$$
(9)

The total surface area, S_t, is assessed from the slope of the low-pressure part of the -plot:

$$S_t = \kappa\frac{a_s^0}{a_s(0.4)}N_A\omega$$
(10)

The difference between the total surface area and the mesopore surface area is often used to estimate the surface area of microspores, :

$$S_{mi} = S_t - S_{me}$$
(11)

The -plot method provides an effective and simple way for evaluation of the micro pore volume V_{mi}, the total surface area S_t and the mesopore surface area same of nonporous materials. The -plots for the active carbons studied obtained by using nitrogen reference adsorption data for non-graphized carbon structures. The slope of the dashed line was used to calculate the total specific surface area, S_t, the intercept of the dotted line was used to evaluate the microspore volume V_{mi}, and the slope of the dotted line was used to evaluate the mesopore surface area S_{me} [1, 2, 121, 145, 146–162].

2.2.2 THEORETICAL MODELS FOR ADSORPTION POTENTIAL DISTRIBUTION CALCULATING

The amount adsorbed A(after conversion to the adsorbate volume V) measures the pore volume accessible to adsorption. If V_t is the maximum volume adsorbed, the difference V_t-V represents the unoccupied pore volume associated with the adsorption potentials smaller than A, and denotes the nonnormalized integral distribution function of the adsorption potential, X*(A). Its first derivative with respect to A is the nonnormalized differential distribution function X (A):

$$X(A) = \frac{dX_n^*(A)}{dA} = \frac{d(V_t - V)}{dA} = \frac{dV(A)}{dA} \qquad (12)$$

In terms of the condensation approximation the adsorption potential distribution (APD), X (A), gives essentially the same information as the distribution function of the adsorption energy. Because for micro porous carbons the amount adsorbed plotted against the adsorption potential is a temperature independent function, the use of Eq. (12) for calculation of APD is fully justified. The distribution function X(A) can be calculated by numerical differentiation of the characteristic adsorption curve V(A), which is obtained by plotting the amount adsorbed as a function of the adsorption potential. Previous experimental and theoretical studies of nitrogen adsorption on active carbons indicate some opportunities for using APD for characterization of these adsorbents. It was shown that the minimum or inflection point on the nitrogen APD curve located at about 4 kJ mol^{-1} before the peak representing the monolayer formation), can be used to evaluate the monolayer capacity. However, the other minimum, which appears in the range of lower adsorption potentials, determines the micro-spore volume. Thus, a typical APD curve for nitrogen on a micro porous carbon at 77 Kexhibits two distinct peaks, which represent the monolayer formation (the peak located between 4 and 8kJ mol^{-1}) and the subsequent volume filling of microspores (the peak between 0.5 and 3 kJ mol^{-1}). The APD curves for nitrogen on the micro and mesopores carbons at 77 K are decreasing functions of A and exhibit an inflection point in the range between 3 and 5kJ/mol, which reflects the completion of the monolayer formation. A sharp increase of the APD curve with decreasing value of A from two to zero reflects the capillary condensation in mesopore. A comparison of the total specific surface area and those evaluated by the BET method and by the -plot method (shows that the APD method gives much smaller values. The values of S$_{t, X(A)}$ seem to be more realistic than those obtained by the BET method and -plot analysis, which are based on the BET model, because in contrast to the BET model the APD method allows for a more accurate estimation of the amount adsorbed in the monolayer. The methods based on the BET model do not take into account the correction for molecules adsorbed inside microspores and therefore, they over estimate the total specific surface area of micro porous solids where the relative adsorption is and is the thermal coefficient ofthe limiting adsorption taken with the minus sign. We could be expressed in terms of the characteristic adsorption curve and the adsorption potential distribution:

$$\Delta S = \left(\frac{\partial A}{\partial T}\right)_{\theta} - \alpha \frac{\theta(A)}{X(A)} \tag{13}$$

This function provides quantitative information about the distribution of the Gibbs free energy for a heterogeneous porous solid and through Eq. (13) it allows one to estimate the entropy and enthalpy [1, 2, 16, 24–37, 146–150].

2.2.2.1   LENNARD–JONES POTENTIAL FUNCTION

In this work, all of the particles include hydrogen molecules, carbon monoxide molecules and carbon atoms are treated as structure less spheres. Particle–particle interactions between them are modeled with Lennard–Jones potential located at the mass-center of the particles. For a pair of particles i and j separated by the distance r, the interaction between them is given by:

$$\phi_{ij}(r) = 4\varepsilon_{ij}\left[\left(\frac{\sigma_{ij}}{r}\right)^{12} - \left(\frac{\sigma_{ij}}{r}\right)^{6}\right] \tag{14}$$

where i and j donate hydrogen, or carbon monoxide, or carbon particles, ε and σ are the energy and size potential parameters, which are 36.7 K and 0.296 nm for hydrogen, 100.2 K and 0.3763 nm for carbon monoxide, and 28.2 K and 0.335 nm for carbon, respectively. Lorentz–Berthelot rules are used to calculate the parameters of interaction between different kinds of particles. The general potential between a gas molecule and the nanotube is calculated by summing up pair interactions between individual carbon atoms and the gas molecules:

$$V(r) = \sum_{j}\phi_{ij}\left(\left|r_i - r_j\right|\right) \tag{15}$$

where r_i is the position of hydrogen or carbon monoxide molecule, r_j is that of a carbon atom and ϕij (r) is the pair potential between a certain gas molecule and a carbon atom, while the sum is over all of the atoms on the tube. Assuming that the atoms of the solid are distributed continuously up and on a sequence of parallel surfaces that form the pore wall, the interaction potential of the fluid molecule with one of these surfaces of area A and number density θ is given by:

$$V(r) = \int_{A}\theta v(r)d\alpha = 4\varepsilon_{FC}\theta\int_{A}\left[\left(\frac{\sigma_{FC}}{r}\right)^{12} - \left(\frac{\sigma_{FC}}{r}\right)^{6}\right]d\alpha \tag{16}$$

where θ = 38 nm^{-2}, ε_{FC} and σ_{FC} are Lennard–Jones parameters of the interaction between fluid molecule and carbon atom. The fluids include hydrogen and carbon monoxide. Integrating over the whole nanotube, the following expression can be obtained:

$$V(r, R) = 3\pi\theta\varepsilon_{FC}\sigma_{FC}^{2}\left[\frac{21}{32}\left(\frac{\sigma_{FC}}{R}\right)^{10}M_{11}(x) - \left(\frac{\sigma_{FC}}{R}\right)^{4}M_{5}(x)\right] \tag{17}$$

where r is the distance between the an atom and the nearest point on the cylinder, R the radius of the nanotube, $x = r/R$ the ratio of distance to radius and θ the same surface number density as above. Here the following integrals are used:

$$M_n(x) = \int_0^\pi \frac{1}{\left(1 + x^2 - 2x \cos \cos \varphi\right)^{n2}} d\varphi \tag{18}$$

Simpson integration is used to get the final potential. For simplicity of programming; the following expression was adopted to represent the results of the integration

$$\frac{V(r,R)}{\varepsilon} = a\left(\frac{\sigma^{10}}{R^{10}}\right) + b\left(\frac{\sigma^4}{R^4}\right) + c\left(\frac{\sigma^4}{R^3}\right) + d\left(\frac{\sigma^4}{R^2}\right) + e\left(\frac{\sigma^4}{R}\right) \tag{19}$$

where a–e are constants dependent on the kind of fluid and the R radius of the nanotubes. They can be determined from the numerical integration. To binary system, the selectivity is defined as:

$$S = \frac{x_1 / y_2}{x_2 / y_2} = \frac{\rho_{p2}^* \rho_{b1}^*}{\rho_{p1}^* \rho_{b2}^*} \tag{20}$$

where x refers to a pore mole fraction and y to a bulk mole fraction. The subscripts of x and y refers to different components in the mixture. is the reduced number density of the bulk fluid. is that of the pore fluid [1, 131, 139, 143, 145, 156, 161, 164].

2.2.3 THEORETICAL MODELS FOR ADSORPTION INTEGRAL CALCULATING

Strongly activated carbons possess a broad distribution of micro pores because some walls between adjacent micro pores burn off. In this case the DR and DA equations cannot give a satisfactory representation of adsorption data. Based on the equations mentioned before that describes adsorption in uniform micro pores, Izotova and Dubinin proposed the following two-term equation for description of adsorption on solids with bimodal micro porous structure:

$$a_{mi} = a_{mi}^{0I} \exp\left[-B_I\left(\frac{A}{\beta}\right)^2\right] + a_{mi}^{0II} \exp\left[-B_{II}\left(\frac{A}{\beta}\right)^2\right] \qquad (21)$$

Here parameters and correspond to micropores, and and correspond to super micro pores. The Eq. (21) has been often applied to describe adsorption on non-uniform micro porous solids. The model studies showed that Eq. (21) is especially suitable for description of adsorption on micro porous solids that possess two types of micro pores of considerably different sizes. For micro porous solids with a great number of micro pores of different sizes, the summation in Eq. (21) should be replaced by integration and then a_{mi} is given by:

$$a_{mi} = a_{mi}^0 \int_0^\infty \exp\left[-B\left(\frac{A}{\beta}\right)^2\right] F(B) dB \qquad (22)$$

where F (B) is the distribution function of the structural parameter B normalized to unity. The integral Eq. (22) was first proposed by Stoeckli. Expressing in this integral the structural parameter B by means of the half-width x:

$$B = \varsigma x^2 \qquad (23)$$

where is the proportionality constant; we have:

$$a_{mi} = a_{mi}^0 \int_0^\infty \exp\left(-mx^2 A^2\right) J(x) dx \qquad (24)$$

where is the micro pore size distribution and m is defined by:

$$m = \frac{\varsigma}{\beta^2} \qquad (25)$$

Comparison of the integrals Eqs. (22) and (24) gives the following relationship between the distribution functions F (B) and J (x):

$$J(x) = 2\varsigma x F\left[B(x)\right] \qquad (26)$$

$$X_{mi}(A) = 2A\beta^{-2}m\int_0^\infty x^2 \exp\left[-mx^2A^2\right]J(x)dx \tag{27}$$

The Eqs. (26) and (27) define the relationship between distribution function $X_{mi}(A)$, F(B) and J(x). The average adsorption potential associated with Eqs. (26) and (27) is given by:

$$\bar{A} = \frac{\beta\sqrt{\pi}}{2}\int_0^\infty \frac{F(B)}{\sqrt{B}}dB = \frac{1}{2}\left(\frac{\pi}{m}\right)^{1/2}\int_0^\infty \frac{J(x)}{x}dx \tag{28}$$

The dispersion for the distribution function $X_{mi}(A)$ may be expressed as follows:

$$\sigma_A = \left[\beta^2\int_0^\infty \frac{F(B)}{B}dB - \bar{A}^{-2}\right]^{1/2} = \left[\frac{1}{m}\int_0^\infty \frac{J(x)}{x}dx - \bar{A}^{-2}\right]^{1/2} \tag{29}$$

where A is defined by Eqs. (28) and (29) have general character and permit calculation of A and for arbitrary micropore distributions F(B) and J(x) [1, 2].

2.3 BASIC MODELS FOR PSD CALCULATING FROM DATA OF SIMULATED ADSORPTION ISOTHERMS

2.3.1 THEORETICAL MODELS

Physical adsorption of gases and vapors on a nonporous surface or on the mesopore surface occurs via layer-by-layer mechanism, whereas adsorption in micro pores resembles the volume filling mechanism. In the case of porous solids containing micro pores and mesopore, active carbons, ordered micro porous and mesopores carbons and active carbon fibers, the volume filling of micro pores occurs first at low pressures and it is followed by the formation of a multilayer film on the mesopore walls, and finally, the remaining empty space inside mesopore is filled via capillary condensation. Thus, the dependence of the amount adsorbed on a porous solid plotted (compared) against the amount adsorbed on reference non-porous solids is linear at higher pressures because the layer-by-layer adsorption occurs on both solid surfaces. However, at low pressures the adsorption mechanisms on the solid studied and on the reference adsorbent can be the same or different, resulting in linear or nonlinear behavior of the initial segment of the comparative plot. It should be noted that the linear segment of the comparative plot at higher pressures is rather insensitive on the choice of the reference solid because after all micro pores are filled and the first adsorbed layer is completed, the surface effects are negligible and the film formation is mostly controlled by adsorbate-adsorbate interactions. The slope of the linear segment at the low pressures is proportional to the total surface area and the slope of the linear segment at high pressures is proportional to the external surface area of mesopore, whereas its intercept determines the maximum amount adsorbed in micro pores, which can be converted to the micro pore volume. There are several types of comparative plots such as the t-plot, -plot and -plot, which differ only in the way of presenting the standard adsorption isotherm measured on the reference solid. In the case of the -plot, the standard isotherm isexpressed in terms of the surface coverage , which is the ratio of the amount adsorbed to the monolayer capacity. The thickness of the surface film on the reference solid,, which is obtained by multiplication of the surface coverage by the monolayer thickness, is used to construct the t-plot. In the -method the amount adsorbed on a porous solid isplotted against the reduced standard adsorption . The -method was proposed. A brief description of this method is provided below. The total amount adsorbed at relative pressure is the sum of the amount adsorbed in the micro pores and the amount adsorbed on the mesopore surface :

$$a\left(\frac{p}{p_0}\right) = a_{mi}^0 \theta_{mi}\left(\frac{p}{p_0}\right) + a_{me}^0 \theta\left(\frac{p}{p_0}\right) \qquad (30)$$

Here, denotes the maximum amount adsorbed in the micropores, and denotes themonolayer capacity of the mesopore surface. For the standard adsorption isotherm measuredon a nonporous reference adsorbent, we have:

$$a_s = \frac{a_s\left(\frac{p}{p_0}\right)}{a_s(0.4)} \qquad (31)$$

Here, is the relative surface coverage for a nonporous reference adsorbent, isthe amount adsorbed on the surface of this reference adsorbent at relative pressure denotes the mono layer capacity evaluated from the standard adsorption isotherm, and denotes the amount adsorbed on the surface of the reference solid at relative pressure and we have:

$$a\left(\frac{p}{p_0}\right) = a_{mi}^0 \theta_{mi}\left(\frac{p}{p_0}\right) + \eta \alpha_s \tag{32}$$

$$\eta = \frac{a_{me}^0 a_s(0.4)}{a_s^0} \tag{33}$$

Kelvin-type equation is usually used to relate the capillary condensation pressure to the radius of cylindrical pores:

$$r\left(\frac{p}{p_0}\right) = \frac{2\gamma V_m}{RT \ln\ln\left(\frac{p_0}{p}\right)} + t\left(\frac{p}{p_0}\right) \tag{34}$$

This comparison involved t-curves proposed by Harkins and Jura:

$$t\left(\frac{p}{p_0}\right) = 0.1\left[\frac{13.99}{0.034 - \log\left(\frac{p_0}{p}\right)}\right]^{0.5} \tag{35}$$

and Halsey model:

$$t\left(\frac{p}{p_0}\right) = 0.354\left[\frac{-5}{\ln\ln\left(\frac{p_0}{p}\right)}\right]^{0.333} \tag{36}$$

where $t(p/p_o)$ is the statistical film thickness (in nm) for nitrogen adsorbed on the carbon surface and p/p_o is the relative pressure. It was shown that the applicability of the Barrett, Joyner and Halenda (BJH) computational method based on the Kelvin equation could be extended significantly towards small mesopore and large micro pores when a proper t-curve was used to represent the film thickness of nitrogen adsorbed on the carbon surface. The t-curve proposed in the work gave the pore-size distribution functions for the carbons studied that reproduce the total pore volume and show realistic behavior in the range at the borderline between micro pores and mesopores [1, 2, 16, 24, 37, 121, 145, 146, 150, 162].

2.3.2 DUBININ-STOECKLI (DS) EQUATION

The adsorption of vapors by micro porous carbons was described by the following fundamental equation of Dubinin-Astakhov (DA):

$$W = W_0 \exp \exp\left[-\left(\frac{A}{E}\right)^n\right] \tag{37}$$

where, W (mmol·g^{-1}) represents the amount adsorbed at relative pressure P_0/P, W_0 denotes the limiting amount of micro pores filling, and A is the differential molar work of adsorption defined as A = RTln (P_0/P) at temperature of T. One may write that E = βE_0, where β is the affinity coefficient depending on the adsorptive only, and it has been assumed that for benzene as a reference β = 1. In general case of heterogeneous micro porous adsorbents, the adsorption is described by the DS adsorption equation of:

$$W = \frac{W_0}{2\sqrt{1+2m\delta^2 A^2}}\exp\left(-\frac{mx_0^2 A^2}{1+2m\delta^2 A^2}\right) \times \left[1 + erf\left(\frac{x_0}{\delta\sqrt{2}\sqrt{1+2m\delta^2 A^2}}\right)\right] \tag{38}$$

which implies a normal half-width (x) distribution of micro pore volume for the slit-like pores as:

$$\frac{dW}{dx} = \frac{W_0}{\delta\sqrt{2\pi}}\exp\left[-\frac{(x-x_0)^2}{2\delta^2}\right] \tag{39}$$

where, x_0 is the half-width of a slit shaped micro pore, which corresponds to the maximum of the distribution curve, and δ is the variance. The letter m is a constant coefficient for a given vapor:

$$m = \left(\frac{1}{\beta k}\right)^2 \tag{40}$$

For benzene as the reference vapor, the constant k equals to 12 kJ·nm·mol^{-1}. Using Eq. (38) to fit the experimental data, three parameters of W_0, x_0 and δ, can be extracted. Knowing these parameters, the micro pore size distribution in terms of volume can be calculated from Eq. (39) [1, 2, 19, 126–133].

2.3.3 STOECKLI MODEL

Another approach for the determination of PSD of porous adsorbents which is also based on the Dubinin's TVFM is Stoeckli method. It had been shown by Stoeckli et al., that for the ideal slit-shaped micro porous materials, a good estimate of the adsorption isotherm can be obtained by:

$$W = W_0 \left[\frac{a}{a + \left(A / \beta K_0 \right)^3} \right] \tag{41}$$

$$K_0 = \frac{10.8 E_0}{\left(E_0 - 11.4 KJmol^{-1} \right)} \tag{42}$$

The a and v are constant parameters that are related to the mean and width of the distribution, respectively. K_0 is calculated using Eq. (42). This is applicable over a range of pore size from 0.4 to 2.0 nm. After obtaining the model parameters using Eq. (41), pore size distribution can be determined using the following gamma type distribution of the mean pore width (L = 2 xs) [1, 126–133]:

$$\frac{dW}{dL} = \frac{3 W_0 a^v L^{3v-1} \exp(-aL^3)}{\tilde{A}(v)} \tag{43}$$

2.3.4 HORVATH-KAWAZOE (HK) METHOD

Horvath and Kawazoe developed a rather simple means of characterizing the pore structure of porous materials. This model provides a simple, one-to-one correspondence between the pore size and relative pressure at which the pore is filled. Using thermodynamic arguments and applying the potential obtained by Horvath and Kawazoe derived the following expression:

$$RT \ln \ln \left(\frac{P}{P_0} \right) = N_{AW} \frac{N_a A_a + N_A A_A}{\delta^4 (L - 2d_0)} \times \left[\frac{\delta^4}{3(L - d_0)^3} - \frac{\delta^{10}}{9(L - d_0)^9} - \frac{\delta^4}{3d_0^3} + \frac{\delta^{10}}{9d_0^9} \right] \tag{44}$$

where L represents the micro pore width (L = 2x), N_{AV} denotes Avogadro's number, and R and T are gas constant and temperature, respectively.

A_a and A_A are dispersion constant characterizing adsorbate-adsorbent and adsorbate interactions, N_a and N_A are the number of atoms per unit area of adsorbent and the number of molecules per unit area of adsorbate, d_0 is the arithmetic mean of the adsorbate molecular diameter and the adsorbent atomic diameter, and δ is the distance between a gas molecule and an adsorbent automat zero interaction energy at relative pressure of P/P_0. From the amount adsorbed at relative pressure of P/P_0, Eq. (44) yields the corresponding slit-pore width, L. Thus, a plot of adsorbed volume versus L is a cumulative pore-volume curve, the slopes of which give the differential PSD [133, 147, 152, 160].

2.3.5 IMPROVED HORVATH-KAWAZOE (IHK) METHOD

The HK equation is widely used for calculating the micro pore size distribution (MPSD) from a single adsorption isotherm measured at subcritical temperature (e.g., N_2 at 77 K). In the HK model, the ideal Henry's law (or linear behavior) is assumed for the isotherm. Cheng and Yang [51] modified the IHK formulation by assuming the nonlinearity of the isotherm equation. This has improved the HK model significantly with the advantage of maintaining the PSD calculation simple. The nonlinearity assumption also results in sharpening PSD. Considering the mentioned nonlinearity correction, researchers derived the IHK for three different pore geometries. For the slit-shaped pores, the IHK equation is derived as:

$$RT \ln \ln \left(\frac{P}{P_0} \right) + \left[RT - \frac{RT}{\theta} \ln \ln \frac{1}{1-\theta} \right] = N_{AV} \frac{N_a A_a + N_A A_A}{\delta^4 (L - 2d_0)}$$

$$\times \left[\frac{\delta^4}{3(L-d_0)^3} - \frac{\delta^{10}}{9(L-d_0)^9} - \frac{\delta^4}{3d_0^3} + \frac{\delta^{10}}{9d_0^9} \right]$$

(45)

The influence of this term and thus θ depends on P/P_0 where the adsorption occurs and also the shape of the isotherm. In the initial part of the adsorption isotherm, where θ is small and θ-dependent term (second term in LHS of IHK equation) approaches zero, the IHK model approaches HK. By increasing θ, the pour filling term $RT - (RT/\theta) \cdot \ln [1/(1-\theta)]$ becomes more negative. In the meantime, as the relative pressure is increased, the free energy term $RT \ln (P/P_0)$ increases. The increase in free energy term is partly offset by the pour filling term. So, the LHS of the IHK equation is increased at a slower rate as compared with the original HK equation.

Consequently, the calculated pore size is increased at a slower rate, result-ing in sharpening the pore size distribution. In this chapter, θ has been calculated from the f DR equation [133, 152, 160].

$$\theta = \frac{C_\mu}{C_{\mu s}} = \exp\left[-\frac{1}{(\beta E_0)^2}\left(R_g T \ln \ln \frac{P}{P_0}\right)^2\right] \tag{46}$$

2.3.6   MOLECULAR SIMULATION POROSIMETRY

The experimental adsorption isotherm measured on a porous solid sample is the aggregate of the isotherms for the individual pores of different sizes. Consequently, the experimental isotherm is the integral of the single pore isotherm multiplied by the pore size distribution, if we neglect the geo-metrical and chemical heterogeneities in the porous surfaces. For a slit-shaped pore, this can be described as:

$$N(P) = \int_{H_{min}}^{H_{max}} f(H)\rho(P,H)dH \tag{47}$$

where N(P) is the amount adsorbed at pressure p, H_{min} and H_{max} are the widths of the smallest and largest pores, (P, H) is the mean density of N_2 at pressure P in a pore of width H. The N (P) versus P relation is just an adsorption isotherm. f(H) is a pore size distribution function, the distribu-tion of pore volumes as a function of pore width H. Therefore, all of the heterogeneities of less crystalline porous solids are approximated by the distribution of pore sizes. If (P, H) can be obtained from the molecular sta-tistics, f (H) can be determined by the best lit to the observed experimental isotherm. The width H in f (H) is not the effective pore width that as men-tioned in above models. In order to derive the molecular density in a pore, statistical approaches to fluids have been used. Seaton et al. applied the mean field theory to calculate (P, H). The mean field theory is an approxi-mate theory of inhomogeneous fluids in which the interactions between the fluid molecules are divided into a short-range, repulsive part and a long-range, attractive part. Each is treated separately for faster calculation than full molecular simulation. The contribution of the long-range forces to the fluid properties is treated in the mean field approximation, while the effect of the short-range forces is modeled by an equivalent array of hard spheres. There are two approaches to get the short-range forces-the local mean field theory and the non-local one, where the former neglects the short-range correlation, bat the latter takes it into account. Seaton et al. adopted the local density approach for their calculation. They calculated (P, H) by the above method. How can we determine f (H) from the calculated (P, H) and the experimental adsorption isotherm N (P)? It has a mathematical difficulty. They used the following bimodal log-normal distribution, which is flexible to represent the various pore size distributions and is zero for all negative pore widths:

$$f(H) = \{V_1/[\sigma H(2\pi)^{1/2}]\}\exp\{-[\log H - \mu_1]^2/2\sigma_1^2\} + \{V_2/[\sigma_2 H(2\pi)^{1/2}]\} \quad (48)$$
$$\times \exp\{-[\log H - \mu_2]^2/2\sigma_2^2\}$$

where V_i is the pore volume of the distribution i, and a, and, u, are the parameters defining the distribution shape. These six parameters in Eq. (48) are determined from the best fit to the experimental adsorption isotherm. The limit of Hmi corresponds to the smallest pour into which the N_2 molecule can enter. On the other hand, the upper limit of H is determined by the width of the mesopore, which condenses at the highest experimental pressure. This calculation can determine the pore size distribution from micro pore to mesopore. In that work the applicability for the pores of less than 1.3 nm was not shown. Scientists extended the above method to the nonlocal mean field theory. The nonlocal mean field theory gives a quantitative accurate description of even ultra micro pore structures. They compared the pore size distributions from the local and nonlocal mean field theories as to real adsorption isotherms by activated carbons; the local theory underestimates the pore size distribution compared with the nonlocal theory. As the calculation with the mean field density theory often gives a qualitative agreement rather than quantitative one. The grand canonical ensemble Monte Carlo simulation is also necessary for such an approach. They also got good results. The molecular simulation studies on the pore size distribution have shown a new picture on the adsorption in the wide range of pores from ultra micro pores to mesopores. Understanding of microspores filling and capillary condensation proceeds rapidly. In the future the pore connectivity will be taken into account, so that a more elaborated method will be settled, although so far molecular simulation porosimetry is not a popular method [1, 2, 162].

2.3.7 ADVANCE METHODS BASED ON DENSITY FUNCTIONAL THEORY

Beside classical methods of pore size analysis, there are many advanced methods. Some researchers proposed a method based on the mean field theory. Initially this method was less accurate in the range of small pore sizes, but even so it gave a more realistic way for evaluation of the pore size distribution than the classical methods based on the Kelvin equation. More rigorous methods based on molecular approaches such as grand canonical Monte Carlo (GCMC) simulations and nonlocal density functional theory (NLDFT) have been developed and their use for pore size analysis of active carbons is continuously growing. Let us consider a one-component fluid confined in a pore of given size and shape, which is itself located within a well-defined solid structure. We suppose that the pore is open and the confined fluid is in thermodynamic equilibrium with the same fluid (gas or liquid) in the bulk state at a given temperature. As the bulk fluid is homogeneous, its chemical potential is simply determined by the pressure and temperature. The fluid in the pore is not of constant density and it is subjected to adsorption forces in the vicinity of the pore walls. This in homogeneous fluid, which is stable under the influence of the external field, is in effect a layer-wise distribution of the adsorbate. The density distribution can be characterized in terms of a density profile, p(r), expressed as a function of distance, r, from the wall across the pore. In the DFT the statistical mechanical grand canonical ensemble is used. The appropriate free energy quantity is the grand Helmholtz free energy, or grand potential functional, D(r). This free energy functional is expressed in terms of the density profile p(r): then by minimizing the free energy (at constant, V, T) it is possible in principle to obtain the equilibrium density profile. For a one-component fluid, which is under the influence of a spatially varying external potential, the grand potential functional, becomes:

$$\Omega\left[\rho(r)\right] = F\left[\rho(r)\right] + \int \rho(r)\left[\Phi(r) - \mu\right]dr \tag{49}$$

where F[p(r)] is the intrinsic Helmholtz free energy functional, is the external potential, and the integration is performed over the pore volume V.

The F[p(r)] functional can be separated into an ideal gas term and contributions from the repulsive and attractive forces between the adsorbed molecules the fluid- fluid interactions) [1, 2, 11, 36, 153, 155].

Hard-sphere repulsion and pair wise Lennard-Jones potential are usually assumed and a mean field treatment is generally applied to the long-range attraction. In the earliest local version of density functional theory (LDFT) the Helmholtz free energy was assumed to be a single valued function of the local density p(r). Further investigations based on comparison of the density profile provided by the LDFT and that determined with GCMC simulations showed that in the case of inhomogeneous fluid more rigorous analysis requires accounting for the density distribution in the region of a few collision diameters in the proximity of a given point. For this reason, it is now customary to apply NLDFT, which involves the incorporation of short-range smoothing functions. In this manner, it has been possible to obtain good agreement with the density profiles determined by Monte Carlo molecular simulations. The NLDFT is well established and widely presented in the literature. The distribution of density in a confined pore can be obtained for an open system in which a pore is allowed to exchange mass with the surroundings. From the thermodynamic principle, the density distribution is obtained by minimization of the following grand potential written below for the one-dimensional case:

$$\Omega\left[\rho(z)\right] = \int \rho(z)\left[f(z) + \Phi(r) - \mu\right]dz \tag{50}$$

where p (z) is the local density of the adsorbed fluid at a distance z from one of the walls of the pore, f (z) is the intrinsic molecular Helmholtz free energy of the adsorbate phase, is the chemical potential. The flee energy f(z) comprises the ideal, mean-field attractive terms, and the excess free energy (repulsive) term as a function of smoothed weighted average. A new approach based on NLDFT to determine pore size distribution (PSD) of active carbons and energetic heterogeneity of the pore wall was proposed by others. The energetic heterogeneity is modeled with an energy distribution function (EDF), describing the distribution of solid- fluid potential well depth (this distribution is a Dirac delta function for an energetically homogeneous surface). The approach allows simultaneous determining of PSD (assuming slit shape) and EDF (from nitrogen isotherms by using a set of local isotherms calculated for a range of pore widths and solid fluid potential well depths. It was found that the structure of the pore wall surface differs significantly from that of graphitized carbon black. This could be attributed to defects in the crystalline structure of the surface, active oxide centers, finite size of the pore walls (in either wall thickness or pore length), and so forth. Those factors depend on the precursors and the process of carbonization and activation and hence provide a fingerprint for each adsorbent. Ustinov and Do approach gives an accurate representation of the experimental adsorption isotherm. The pore size distributions indicate quite significant differences in the porosity of the carbons studied in the range of micro pores and mesopore [1, 11, 36, 153, 155].

2.3.8   JARONIEC-CHOMA METHOD

The integral equation that mention before was solved for various continuous functions representing the distribution F (B). For example Wojsz and Rozwadowski solved the integral equation for distribution functions F (B) other than the Gaussian one, decreasing and increasing exponential distributions and Rayleigh distribution. Some of these equations have rather complex mathematical form; however, the decreasing exponential and Rayleigh distributions may be considered as special cases of the gamma-type distribution, which generate a very simple isotherm equation. General integral equation similar to integral equation for the adsorption isotherm on heterogeneous micro porous solids can be written as follows:

$$a_{mi} = a_{mi}^0 \int_0^\infty \theta_{mi}(z, A) F(z) dz \qquad (51)$$

where a_{mi} is the equilibrium amount adsorbed in microspores, is the maximum amount adsorbed in micropores, is the local isotherm describing adsorption in uniform micropores; z is the quantity associated with the micropore size; $A = RT \ln(p/p_o)$ is the adsorption potential, F(z) is the distribution function characterizing heterogeneity of micro porous structure. Jaroniec and co-workers proposed the following gamma-type distribution function:

$$F(z) = \left[\frac{n\rho^\nu}{\Gamma\left(\frac{\nu}{n}\right)} \right] z^{\nu-1} \exp\left[-(\rho z)^n \right] \qquad (52)$$

whereis the inverse value of the characteristic energy E_o for the reference adsorbate, $p > 0$ and $v > 0$ are parameters for the gamma distribution function. It was shown elsewhere that the Jaroniec-Choma (JC) equation, which was obtained by generalization of the DA equation for $n = 2$ or $n = 3$, gives good description of gas and vapor adsorption for many micro porous active carbons. A general form of the JC equation can be written as:

$$a_{mi} = a_{mi}^0 \left[1 + \left(\frac{A}{\beta \rho} \right)^n \right]^{-v/n} \tag{53}$$

Here a_{mi} and denote respectively the amount and the maximum amount adsorbed in micro pores, p and v are parameters of the gamma distribution function. Equation(53) with $n = 2$ was proposed by Jaroniec and Choma on the basis of the assumption that mention before with $n = 2$ [Dubinin-Radushkevich (DR) equation] governs adsorption in uniform micro pores. This assumption was justified experimentally by many researchers. Later, Scientists carried out careful adsorption and calorimetric experiments for benzene on molecular carbon sieves, and showed that DA equation with n = 3 gives a better representation of adsorption in uniform micro pores than that with $n = 2$. These experimental studies suggest that the DA equation with $n = 2$ describes adsorption in nearly uniform micro pores and it's generalized from [Eq. (53) with $n = 2$] gives a good description of adsorption on micro porous active carbons with large structural heterogeneity. For micro porous solids with moderate structural heterogeneity, the use of the isotherm equations obtained by generalization of DA equation with $n = 3$ [the JC Eq. (53) with $n = 3$] is substantiated better than the use of those generated by DA equation with $n = 2$. It is noteworthy that some authors postulated that DA equation can be used to describe adsorption in uniform micro pores several years before this postulate found some experimental justification. The distribution function F(z) together with the quantities and are used to characterize the structural heterogeneity of micro porous solids. Energetic heterogeneity of a micro porous solid generated by the overlapping of adsorption forces from the opposite micro pore walls can be described by the adsorption potential distribution in micro pores . This distribution associated with Eq. (53) is given by:

$$X_{mi}(A) = v(\beta \rho)^{-n} A^{n-1} \left[1 + \left(\frac{A}{\beta \rho} \right)^n \right]^{-\frac{v}{n}-1} \tag{54}$$

The adsorption potential distribution in micropores X_{mi}(A) given by Eq. (54) can be characterized by the following quantities. Although description of micro porous structures of nano porous carbons is a difficult and still not fully solved task, comparative studies of various adsorption models can facilitate elaboration of methods for characterization of micro porous solids. It was shown that the gamma distribution function F(z) gives a good description of structural heterogeneity for many micro porous carbonaceous materials. For micro porous active carbons with small structural heterogeneity the JC equation gives a good description of adsorption in micro pores, while the JC equation can be used for adsorption on micro porous active carbons with strong structural heterogeneity [1, 2–19].

2.3.9   DUBININ-RADUSHKEVICH EQUATION

The DR equation, proposed in 1947, undoubtedly occupies a central position in the theory of physical adsorption of gases and vapors on micro porous solids. The amount adsorbed in micro pores a_{mi} is:

$$a_{mi} = a - a_{me} \qquad (55)$$

where a is the sum of the amount adsorbed in micropores a_{mi} and in mesopores ame. According to the DR equation a_{mi} is a simple exponential function of the square of the adsorption potential A:

$$a_{mi} = a_{mi}^0 \exp\left[-B\left(\frac{A}{\beta}\right)^2\right] \qquad (56)$$

Here B is the temperature-independent structural parameter associated with the micropore sizes, and fl is the similarity coefficient, which reflects the adsorbate properties that such equations is commonly used for description of gas and vapor adsorption on micro porous active carbons.

A more general expression is that proposed by Dubinin and Astakhov, which is known as the DA equation: a_{mi}:

$$a_{mi} = a_{mi}^0 \exp\left[-\left(\frac{A}{\beta E_0}\right)^n\right] \qquad (57)$$

In Eqs. (55), (56) and (57) a_{mi} represents the amount adsorbed in micro pores at relative pressure p/p_o and temperature T, is the limiting volume of adsorption or the volume of micropores and Vm is the molar volume of the adsorbate. The specific parameters of Eqs. (55) and (56) are and n, respectively, where E_0 is the characteristic energy of adsorption for the reference vapor, usually benzene. Dubinin proposed to extract the amount a_{mi} adsorbed in micro pores from the total adsorbed amount a (p/p_o) as follows:

$$a_{mi} = a - S_{me}\gamma_s \tag{58}$$

The amount adsorbed per unit surface area was evaluated from the adsorption isotherm measured on a reference adsorbent, whereas the specific surface area Sine of mesopores was estimated from the adsorption isotherm. To analyze nitrogen adsorption isotherms by means of the DR and DA equations, we extracted according to Eq. (58) the adsorption isotherm a_{mi} (p/p_o) for micro pores from the total adsorption isotherm a (p/p_o). To calculate a_{mi} according to Eq. (58), we evaluatedfrom the standard nitrogen isotherm at 77 K shown by the -plot method. The extracted adsorption isotherm was described by DR equation (55) and also by DA equation (56) [1, 19, 126, 150, 152].

2.3.10 DFT- NLDFT MODELS

The PSD is calculated from the experimental adsorption isotherm with (P/P_0) by solving the integral adsorption equation:

$$N_{exp}(P/P_0) = \int_{D_{min}}^{D_{max}} N_{QSDFT}(P/P_0, D) f(D) dD \tag{59}$$

The experimental isotherm is represented as the convolution of the DFT kernel (set of the theoretical isotherms NQSDFT(P/P_0, D)in a series of pores within a given range of pore sizes D) and the unknown PSD function f (D), where D_{min} and D_{max} are the minimum and maximum pore sizes in the kernel, respectively. Two kernels of the selected DFT adsorption isotherms for the slit geometry are reviewed, include NLDFT and QSDFT. In contrast to the NLDFT kernels, the QSDFT isotherms are smooth prior to the capillary condensation step, which is characteristic of mesopore (D > 2 nm), and thus do not exhibit stepwise inflections caused by artificial layering transitions Solution of above equation can be obtained using the quick nonnegative least square method. It should be noted that in NLDFT kernels, the pore width is defined as the center-to-center distance between the outer layers of adsorption centers on the opposite pore walls corrected for the solid–fluid LJ diameter. In QSDFT kernels, the pore width is defined from the condition of zero solid density excess. These definitions are apparently different, albeit insignificantly, but this difference should be taken into account in data analysis. Over the years a library of NLDFT and more recently QSDFT kernels were developed for calculating pore size distributions in carbonaceous and silica micromesoporous materials of different origin from nitrogen and argon adsorption isotherms, as well as for micro porous carbons from carbon dioxide adsorption. For a DFT kernel of a given adsorbate–adsorbent pair, the parameters should not only represent the specifics of adsorbent–adsorbate interactions, but also take into account the pore structure morphology. DFT kernels were built for calculating PSD using different adsorbates: nitrogen at 77.4 K, argon for 77.4 and 87.3 K, and carbon dioxide at 273 K. It was shown, that the results obtained with different adsorbents are in reasonable agreement. Nitrogen at77.4 K is the conventional adsorbate for adsorption characterization [1, 15, 39].

However, argon and carbon dioxide are more suitable in some cases, especially for micro porous materials. In general, argon at87.3^K is always a better molecular probe than nitrogen, since it does not give rise to specific interactions with a variety of surface functional groups, which can lead to enhanced adsorption/specific interactions caused by quadruple moment characteristic to non symmetric molecules. In addition, argon at 87.3 K fills micro pores of dimensions 0.5–1 nm at higher relative pressures compared to nitrogen at 77.4 K, and, due to faster diffusion, the equilibration times are shorter. As such, it is possible to test micro pores as small as 0.5 nm with argon within the reliable range of relative pressures that is limited in modern automated instruments. The advantages of using argon are very pronounced for zealots and metal-organic frameworks. Historically, the first DFT kernels were developed for carbon slit pores some researchers designed the consistent equilibrium NLDFT kernels for nitrogen, argon, carbon dioxide isotherms, which are applicable for disordered micromesoporous carbons of various origin, including activated carbons, and carbon fibers, charcoal, and carbon black. Activated carbon fibers (ACF)exhibit a type I isotherm and possess a very high adsorption capacity with BET surface areas up to 3000 m²/g. This results in rapid adsorption and desorption rates with over 90% of the total surface area belonging to micro pores of 2 nm or less. And like their granular counterparts, ACF are finding a foot hole in a broad range of applications including gas and liquid phase adsorption, carbon molecular sieves, catalysis, gas storage, and super capacitors. The factors that greatly affect the ACF properties (precursor source, temperature, time, gas flow activating agents and the use of catalysts) are the ones that most influence the pore structure. ACF prepared by a physical activation process will be dependent on a controlled gasification process at temperatures ranging from 800° 1000 °C. In their activation procedure they applied the NLDFT method and showed that a greater degree of activation led to a widening of the pore size distribution from 2.8 to 7.0 nm. They contribute this broadening to a decrease in the number of micro domains. This phenomenon was coupled with an increase in the peak pore size (from 0.44 nm to 1.86). The adsorption data and subsequential pore size analysis was confirmed by NMR. The chemical activation process on the other hand involves the mixing of a carbon precursor with a chemical activating agent typically KOH, NaOH, H_3PO_4 or $ZnCl_2$ [1, 127, 129, 142, 147, 148, 153, 155].

2.3.11 ADSORPTION INDUCED MOLECULAR TRAPPING (AIMT) IN MICROPOROUS MEMBRANE MODEL

Model and simulation schemes and Principles of the DCV-GCMD numerical experiment in micro porous membrane is illustrated in Fig. 2.1.

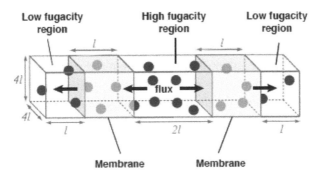

FIGURE 2.1   Schematic representation of DCV-GCMD method.

In our simulations the membrane thickness stands as one unit cell: 1 = 2:5 nm. Our numerical experiment consists in reproducing an experimental set-up used for permeability measurements, as illustrated in below detailed. For that purpose we use the DCV-GCMD method with high and low fugacity reservoirs are imposed at each end of the membranes. This allows the application of periodic boundary conditions. While the fugacity's in these reservoirs are controlled by means of Grand Canonical Monte Carlo simulation, molecular motions are described using Molecular Dynamics simulation. Once the system has reached the steady-state regime, the molar flux J is estimated by counting the number of molecules N crossing the membrane of cross section S during a time interval :

$$J = \frac{N}{S \Delta t} \qquad (60)$$

The permeability is then defined as the flux per unit of fugacity gradient $(D_f = l)$ across the membrane:

$$P = \frac{Jl}{\Delta f} \qquad (61)$$

From the thermo dynamical point of view, a system of fluid particles sorted in a immobile porous medium deviates from equilibrium when a gradient in molecules chemical potential exists. Under isothermal conditions and on the assumption that transport mechanism is diffusive, the local molar flux ~j (number of moles of fluid per unit surface per unit time) through the fixed porous solid satisfies the Maxwell-Stefan equation:

$$\vec{j} = -\frac{cD_o}{RT} \vec{\nabla} \mu \qquad (62)$$

where T is the temperature, R the ideal gas constant, c the average interstitial concentration(number of moles per material unit volume), while Do stands as the collective diffusivity of the sorted fluid, as previously discussed. Moreover, using the definition of the chemical potential:

$$\mu(f, T) \equiv \mu_0(T) + RT \ln \ln(f / f_0) \qquad (63)$$

One can alternatively consider the gradient in fugacity as the driving force of fluid motion. Therefore, the rearranged expression of the local molar flux:

$$\vec{j} = -\frac{cD_o}{f}\vec{\nabla}f \qquad (64)$$

where f is the fluid fugacity. In order to estimate the concentration in the micro porous membrane, refer to the classical Langmuir model, commonly used to describe adsorption isotherms of fluids in micro porous adsorbents:

$$c = c_s\frac{bf}{1+bf} \qquad (65)$$

where is the complete filling concentration and is an equilibrium adsorption constant, which can be interpreted as the inverse of a characteristic filling pressure. We stress that we use the Langmuir model for its ability to reproduce the adsorption isotherms simulated in our membrane models (see supplementary information) and its convenient analytical form. Let us now consider a micro porous membrane of thickness in the direction and separating two infinite bulk fluid reservoirs exhibiting a difference in chemical potentials. Under these conditions, Eq. (65) describes the local motion of interstitial fluid in the membrane. In an actual experiment, one can only measure the total molar flux as a function of the fugacity drop across the membrane. Therefore, the averaged transportequation, obtained from the average of this equation along the thickness of the membrane:

$$\vec{j} = -P_e\frac{\Delta f}{l}\vec{e}_x = -\frac{1}{l}\int_{x=0}^{x=1}dx\frac{D_oc}{f}\left(\frac{\partial f}{\partial x}\right)\vec{e}_x \qquad (66)$$

in which is referred to as the permeability of the membrane. It should be stressed that this definition of the permeability differs from the classical definition deriving from Darcy's law, which considers the viscous flow of a Newtonian interstitial fluid. In the present case, for the sake of generality we define the permeability as the transport coefficient relating the molar flux to the driving force of fugacity gradient, as found in the literature, assuming a constant , we deduce the overall permeability of the membrane from Eq. (67) as:

$$P_e = \frac{D_o}{\Delta f} \int_{f_u}^{f_d} df \frac{c}{f} \tag{67}$$

in which and are the downstream and upstream fugacities respectively. Finally, combining this equations, the permeability is given as a function of and :

$$P_e = \frac{D_o c_s}{\Delta f} \ln\left(1 + \frac{b\Delta f}{1+bf_d}\right) \tag{68}[166].$$

2.3.12   DERJAGUIN-BROEKHOFF-DEBOER MODEL

An improvement of the classical DBD theory for capillary condensation/ evaporation in open-ended cylindrical capillaries was presented in Ref. [42]. Here we reintroduce the main ideas of the DBD theory and present its extension for the capillary condensation/evaporation in spherical mesopores. It was previously shown that the experimental adsorption data for a reference flat silica surface can be properly described by using the disjoining pressure isotherm in the equation:

$$\Pi_1 \exp\left(-h / \lambda_1\right) + \Pi_2 \exp\left(-h / \lambda_2\right) = -\left(RT / v_m\right)\ln\ln\left(p / p_0\right) \tag{69}$$

in which and characterize the strength of the surface forces field, whereas the parameters and are responsible for the range of the structural forces action. Clearly, the first term dominates in thick adlayers, whereas the second term dominates in thin ad layers. All of the parameters appearing in Eq. (69) were tabulated previously for the adsorption of argon and nitrogen at their boiling points on the selected reference silica surface. The critical radius at which a spontaneous capillary condensation occurs (spinodal condensation point) is closely related to the assumed pore geometry. For the wetting films formed on a concave surface of spherical pores, the following relationship is valid:

$$\Pi(h) = \Pi_1 \exp\left(-\frac{h}{\lambda_1}\right) + \Pi_2 \exp\left(-\frac{h}{\lambda_2}\right) + \frac{2\gamma(r_m)}{r-h} = -\left(RT / v_m\right)\ln\ln\left(p / p_0\right) \tag{70}$$

As demonstrated previously, the surface tension of a liquid adsorbate depends on the meniscus radii, which seems to be particularly important for the pores at the borderline between micropores and mesopores. Similar to the work completed by many researchers, the GTKB was used in this chapter. The GTKB equation for the cylindrical interface (capillary condensation) can be written as follows:

$$\frac{\gamma(r_m)}{\gamma_\infty} = 1 - \frac{\delta}{r_m} \tag{71}$$

In Eqs. (71), denotes the surface tension of the bulk fluid, and is the displacement of the surface at zero mass density relative to the tension surface. The physical meaning of and its impact on the spinodal condensation point was presented previously. The stability condition of the wetting film was formulated earlier by researchers as . Obviously, Capillary Condensation/Evaporation in Spherical Cavities both the critical film thickness, and the critical capillary radius, , corresponding to the film collapse, are determined from Eq. (42).

$$\frac{d\Pi(h)}{dh}\Big|_{h=h_{cr}, r=r_{cr}} = 0 \tag{72}$$

The condition given by Eq. (72) determines the spontaneous spinodal condensation when the adsorbed film thickness becomes mechanically unstable.

Combination of Eqs. (69)–(72) gives the relation between both the critical film thickness and the critical capillary radius as a function of the relative pressure for the spherical pore geometry: The solution of this system of algebraic equations can be obtained by chord or other standard numerical procedures. The thickness of the adsorbed film in equilibrium with the meniscus for the spherical pore is given by:

$$RT \ln\ln(p_0 / p) = \Pi(h_e)v_m + \frac{2v_m\gamma_\infty\left(1 - \dfrac{\delta}{r - h_e}\right)}{r - h_e} \tag{73}$$

For constant surface tension Eqs. (72) and (73) reduce to the classical DBD equations, which are described in the series of papers published by Ravikovitch and Neimark [167]. For application purposes, we can derive the analytical formulas for the calculation of the equilibrium transition in the considered spherical pore geometry according this equation. As we mentioned above, for the spherical pore, the system of equations describing the equilibrium desorption transition is defined by the following relation:

$$\theta_{exp}(h) = \int\limits_{\Omega(D)} \theta_{loc}(D_{in}, h) \chi(D_{in}) dD_{in} \tag{74}$$

where,

$$\theta_{loc}(D_{in}, h) = \begin{Bmatrix} t/t_{cr}, t < t_{cr} \\ 1, elsewhere \end{Bmatrix} \tag{75}$$

The Eq. (75) represents the kernel of the theoretical isotherms generated from the DBD approach in pores of different diameters, , and is the normalized differential PSD function. As mentioned above, the kernel of the integral equation (given by Eq. (74)) was obtained assuming there is spinodal condensation in the spherical pores. By applying (as proposed by us previously) the ASA algorithm along with the stabilizing first-order Tihkonov's regularization term and regularization parameter was selected through a series of trials by an interactive judgment of the solution. During the regularization, some of the artificial shoulders and stairs were smoothed out [167].

2.3.13   DRS AND DI MODEL (DUBBINI-IZOTOVA, DUBBINI-RADUSKEVICH-STOEKLI)

The classical Dubinin–Radushkevich (DR) equation is commonly used in its linear form for analysis of active carbons and micro pore size analysis that mention before. The DR equation was applied to the nitrogen experimental isotherm at 77 K using values. From this data, the micropore width (L) and following equations were used with the assumption of a slit-shaped pore:

$$\log(W) = \log(W_0) + M \times \log^2(P_0 / P) \tag{76}$$

The micro pore volume is calculated from the intercept of a log(W) vs. plot, while the slope, M, of the best fit line is related to the adsorption energy, , as follows:

$$M = -2.303 \times (RT / E)^2 \tag{77}$$

where the ideal is gas constant and is the adsorption temperature in Kelvin. It has also been shown that the parameter is related to the average micropore half-width by the equation:

$$x = \frac{\beta k}{E} \tag{78}$$

The similarity coefficient is a shifting factor, which at a given temperature depends only on the adsorbate and is equal to , where is the adsorption energy of a reference vapor (typically benzene). For nitrogen at 77 K, is equal to a value of 0.33 and is a structural parameter that is equal to 13 nm kJ/mol for this set of materials. The micro pore width, is calculated as twice the micro pore half-width. The Dubinin–Radushkevich–Stoeckli (DRS) equation was used to determine the pore size distributions for several of the ACFs. Previously, Daley and co-workers showed a good correlation between this theory and direct measurement of the pore size distributions using STM [176].

The fitting of simulation data by the studied models was performed using the genetic algorithm of simulation. All results were described (in the whole pressure range) by a classical Dubinin–Astakhov adsorption isotherm equation, using the values of the affinity coefficient tabulated in this review paper. Moreover, we applied the Dubinin–Izotova model and the Dubinin–Raduskevich–Stoeckli equation (also in the whole pressure range) in the form:

$$N_{DRS} = \frac{N_{mDRS}}{(1 + \mathrm{erf}(\frac{x_0}{\Delta\sqrt{2}}))\sqrt{1 + 2m\Delta^2 A_{pot}^2}} \times \exp\left[-\frac{A_{pot}^2 m x_0^2}{1 + 2m\Delta^2 A_{pot}^2} \right]$$

$$\times \left[1 + \mathrm{erf}\left(\frac{x_0}{\Delta\sqrt{2}\sqrt{1 + 2m\Delta^2 A_{pot}^2}} \right) \right] \tag{79}$$

where and are the values of adsorption and maximum adsorption, respectively, is the adsorption potential, is a proportional coefficient (is assumed as equal to 12 (kJ nm mol^{-1})), is the similarity coefficient, erf is the error function, and are 'dispersion' and mean of Gaussian distribution, respectively. The pore size distribution was calculated using the correct normalization factor (i.e., from 0 up to):

$$\chi_{normDRS} = \frac{2}{\Delta\sqrt{2}\pi(1+erf\left(\frac{x_0}{\Delta\sqrt{2}}\right))}. \tag{80}$$

Finally, the data were also described using the model proposed by Jaroniec and Choma:

$$\chi_{JCh}(x) = \chi_{normJCh} \exp\left[-\rho\zeta x^n\right] \tag{81}$$

And

$$\chi_{normJCh} = \frac{n(\rho\zeta)^{v+1}}{\Gamma(v+1)} \tag{82}$$

where is the Euler gamma function, is constant equal to and are the parameters of Eqs. (80)–(82). The average micro pore diameters from DI model were calculated using:

$$H_{eff,av,DI} = \frac{N_{m1}H_{eff,av1} + N_{m2}H_{eff,av2}}{N_{m1} + N_{m2}} \tag{83}$$

For the remaining models the average micro pore diameters were calculated from integration of the PSD curve [170].

2.3.14   AVERAGE MICROSPORE DIAMETER (L_{AV}), CHARACTERIZATION ENERGY (E_0) CALCULATION AND ADSORPTION CALORIMETRIC DATA

Relation between average microspore diameter (L_{av}) and characterization energy (E_0), is presented by Eq.(84):

$$L_{av} = \frac{\kappa}{E_0} \qquad (84)$$

where   is the characteristic constant for a defined adsorbate/adsorbent pair in the micro pore region. The value of this characteristic constant for benzene vapor on activated carbon is about 12 kJ nm mol^{-1}. It has been also assumed that this parameter is in a small degree dependent on the characteristic energy of adsorption:

$$L_{av} = \frac{13.028 - 1.53 \times 10^{-5} E_0^{3.5}}{E_0}. \qquad (85)$$

Next, others relations were developed since this turning point. The intensive investigations were done by McEnaney:

$$L_{av} = 6.6 - 1.79 \ln \ln (E_0), \qquad (86)$$

Stoeckli et al., model:

$$L_{av} = \frac{10.8}{E_0 - 11.4}, \qquad (87)$$

$$L_{av} = \left(\frac{30}{E_0}\right) + \left(\frac{5705}{E_0^3}\right) + 0.028 E_0 - 1.49, \qquad (88)$$

Choma and Jaroniec model:

$$L_{av} = \left(\frac{10.416}{E_0}\right) + \left(\frac{13.404}{E_0^3}\right) + 0.008212 E_0 + 0.5114, \qquad (89)$$

and Ohkubo et al.:

$$\ln(p / p_s) = \frac{A_{HK}}{L - d}\left[\frac{B_{HK}}{(L - d/2)^3} - \frac{C_{HK}}{\left(L - \frac{d}{2}\right)^9} - D_{HK}\right] \qquad (90)$$

where   is the sum of the diameter of an adsorbent (d_a) atom and an adsorbate molecule (d_A) (the remaining parameters of the above equation are defined in list of abbreviation). Finally, the following two equations were obtained:

$$\frac{L_{av(all)}}{d_A} = \frac{a_{1(all)}}{1 + b_{1(all)} \exp\left[-c_{1(all)}n\right]} + \frac{a_{2(all)} + b_{2(all)} \times n + c_{2(all)}n^2}{E_0}, \tag{91}$$

Not only for the micro pore had diameters (called the first range):

$$\frac{L_{av(mic)}}{d_A} = \left(\frac{a_{1(mic)}n}{b_{1(mic)} + n}\right) + \left(\frac{a_{2(mic)}n}{b_{2(mic)} + n}\right)^{E_0} \times E_0^{\left(\frac{a_{3(mic)}n}{b_{3(mic)} + n}\right)} \tag{92}$$

For calculations up to effective diameter equal to 2 nm (called the second range). The additional details of all above-mentioned calculations were given previously therefore they are omitted in the current study. Taking into account the assumptions made during the derivation of Eq. (90), the main condition, which should be fulfilled for the chosen molecules of adsorbents, is the absence of a dipole moment and a spherical-like structure. Therefore, with chosen N_2, Ar, CCl_4, and C_6H_6, for some of these adsorbents the Horvath–Kawazoe method was adopted. Moreover, these adsorbents have been widely applied in the investigation of the structural heterogeneity of micro porous carbons. However, the choice of nitrogen at its liquid temperature as the probe molecule may not be suitable for very narrow pores such as those in carbon molecular sieves, where the activated diffusion effects might be important. These effects can be reduced by conducting the experiment at higher temperatures. It is thus useful to investigate the MSD obtained from adsorption of different adsorbents at temperature other than the boiling point of nitrogen at 77.5K, for example at near ambient temperatures. The temperatures chosen in the calculation were equal or very close to those applied in measurements where the investigated adsorbents are used for the determination of structural heterogeneity of carbons. Summing up, all the above results show that the average pore diameter is a function not only of E^{-0} but also of n. Then using Eqs. (91) and/or (92), the average reduced effective diameter can be calculated and multiplied by adsorbate diameter. The typical plots (the lines) for chosen adsorbents (C_6H_6 and N_2) and values of (1.50 and 3.25) are compared with relationships proposed by the other others works. It should be pointed out that the shape of this curve is similar to that observed for the empirical and/

or semiempirical relationships. This procedure was also applied previously to experimental data of adsorption on different carbonaceous molecular sieves. The correlation between suggested and calculated using Eqs. (91) and (92) pore diameters is very good. On the other hand, we recently tried to answer the most general questions [133, 152]. In addition, to perform the thermodynamic verification for both samples studied in this chapter is:

$$q^{diff} = \Delta G^{ads} - \Delta H^{vap} \tag{93}$$

whereand is the enthalpy of vaporization (equal with minus sign to the enthalpy of condensation,). Horvath and Kawazoe noticed the similarity of the data calculated based on Eq. (93) (with the "experimental" esoteric heat of adsorption obtained from isotherms measured at different temperatures. It should be pointed out that the HK model has been verified only for one set of experimental data and we do not find other cases in the literature. Therefore, the calorimetrically measured enthalpies of adsorption of C_6H_6 and CCl_4 are shown in Eqs. (94) and (95). Moreover, these data will be applied to calculate, using the standard method the enhancement of potential energy in micro pores in comparison to the energy of adsorption on a "flat" surface. Knowing adsorption isotherms and the differential molar heats of adsorption, the differential molar entropies of adsorbed molecules () can be calculated by:

$$S^{diff} = S_g - \left(\frac{q^{diff}}{T}\right) - R\ln\ln\left(\frac{p}{p_0}\right) + R, \tag{94}$$

where is the molar entropy of the gas at the temperature and is the standard state pressure. It is well known that different standard states can be chosen, and in our case, the gas at the standard pressure of = 101,325 Pa was applied. Also some researchers calculated the enthalpy and entropy of adsorption basing on the potential theory and applying the procedure described previously. Assuming the fulfillment of the main condition of the potential theory (first of all the temperature invariance conditions that obtained:

$$q^{diff} = A_{pot} + \frac{\alpha T\Theta}{F\left(A_{pot}\right)} + L - RT = A_{pot} + \frac{\alpha T \beta E_0}{n}\left(\frac{A_{pot}}{\beta E_0}\right)^{1-n} + L - RT \quad (95)[152]$$

2.3.15   BIMODAL PORE SIZE DISTRIBUTIONS FOR CARBON STRUCTURES

Every puzzling result are obtained by analyzing micro porosity of various carbonaceous materials of different origin and/or treated thermally or chemically since the difference between the micro pore size distributions curves are often in significant. The number of peaks on the pore size distribution (PSD) curves and the ranges of their location as well as their shapes are very similar. Thus, the bimodality of PSD for many carbonaceous adsorbents is well known fact in the literature. This fact can be explained by conditions of carbon preparation; for example, by creation of the porous structure during activation process. It is amazing that different carbon materials obtained from different precursors have similar (still bimodal) porous structure with the gap between the peaks reflecting to the filling of primary and secondary micro pores. Since the recent recipes for the preparation and manufacturing of new carbonaceous adsorbents are still developed, the studies of the methods for porosity characterization should be improved and this improvement should enhance the method sensitivity. It is not surprising that the above topics attract a lot of attention. Moreover a systematic numerical investigation of the effect of bimodality of PSD and its sensitivity on the reconstruction has not been published yet. Some researchers observed that the micro pore size distributions determined from experimental isotherms (GCMC and DFT) usually show minima near two and three molecular diameters of the effective pore width (1 nm), regardless of the simulation method used. The proposed explanation was that this is a model-induced artifact arising from the strong packing effects exhibited by a parallel wall model. Moreover, the inclusion of surface heterogeneity in the DFT model used to generate the local adsorption isotherms (while more realistic) did not change this observation significantly. Many researchers argued that these artificial minima are primarily due to the homogeneous nature of the model (a sharp monolayer formation occurring at approximately the same relative pressure in most of the pores, and the second layer existence in pores between 1.0 and 2.0 nm) that can accommodate four or more layers forming at the relative pressure of ca. 0.1. In the other words, since all theoretical isotherms in wide pores exhibit a monolayer step at about the same pressure, the contribution from pores that fill at this pressure would have to be reduced. This compensation effect is responsible for the observed minima on the pore size distribution curves, and also for the deviations between best fit and experimental isotherms. Similarly, less pronounced minima occur for pores, which fill at the relative pressures corresponding to the formation of the second, third, and higher layers. In addition, many scientists investigated the effect of heterogeneity of the pore walls (differences in the pore wall thicknesses) on the PSD shape. They pointed out that the influence of the pore wall thickness is significant, especially for lower values of relative pressures since the inter molecular potential is dominated by interactions with this wall. Consequently, the pore size distributions are shifted to smaller pores as the pore wall thickness decreases. However, the minima at 1 and 2 nm were observed for all systems studied. The impact of the boundary values of nitrogen relative pressure (p/ps) was analyzed on the basis of the observed alterations in PSDs by others [169].

They observed that the changes in micro porosity and mesoporosity of activated carbons can be described adequately only when the range of p/p_s is as wide as possible. The PSD curves can be broadened with shifted maxima especially for micro pores and narrow mesopores when adsorption data start at relatively high p/ps value. However, the differences between the respective pore size distribution functions are in significant. The influence of a random noise on the stability of the solution of the integral equation (calculation of PSD) was also studied on the basis of theoretically generated isotherms. For low and medium noise levels, the reproducibility of some perturbed isotherms was good. Summing up, the "true" and some perturbed generated isotherms (also experimental ones) contain full information about the assumed distribution (number of peaks, their location, area, etc.). Moreover, it was shown that an increase of the smoothing (regularization, the INTEG algorithm) parameter leads to a strong smoothness and disappearance of some (reasonable for experimental studies) peaks on the distribution curve (i.e., the PSD shape changes from initial poly modal distribution to much flatter mono modal one). It should be pointed out that similar results were obtained by others. In recent studies some systematic investigations of the influence of pore structure and adsorbate-adsorbate and adsorbate-adsorbent parameters on the pore size distribution functions were performed. Thus, it was shown on the basis of experimental and theoretical data that the changes in the shape and behavior of the local adsorption isotherm do not guarantee the differences between the evaluated PSDs except the strictly micro porous adsorbents. All the observations suggested that for some parameters the larger pore diameters of the PSD function are, the smaller changes in the value of average pore diameter were observed. Moreover, the influence of the energetic heterogeneity on the structural parameters and quantities is rather insignificant; all calculations led to almost the same PSD curves and this similarity was observed for the adsorbents of different origin and possessing different pore structures. In conclusion, the following question arises: Why and when the gap between peaks of the differential pore size distribution curves can be related to the mechanism of the primary and secondary microspore filling? In order to answer this question we will use the experimental data published by some researchers showing the development of secondary porosity in a series of carbons. These data will be described by applying the proposed previously the ASA algorithm with the method of Nguyen and Do (ND). Next it will be shown that the behavior of the experimental systems studied represents a specific case, which can be, together with other systems, the subject of the proposed general analysis of the behavior of PSD curves. Therefore, by using carbon samples with varying porous structure it is possible to perform a systematic study of various situations related to the shape of the PSD curves, that is, the intensity of the both peaks, their mutual location and the vanishing of one of them. It is also possible to obtain the information how far the bimodality of those distributions is retained and reconstructed. Moreover, the problem in the similarity of the local adsorption isotherms generated for the range of pore widths corresponding to the gap between peaks will be discussed way, that is, the PSD curves (J (Heff)) using the ASA algorithm and the method proposed by Do and co-workers are calculated. It should be noted that 82 local isotherms generated for the same effective widths changing from 0.465 up to 233.9 nm were used. Additional details of the preparation of carbons and the procedure of the differential PSD calculations were given previously. Nearly all distribution curves show the existence of the bimodal porous structure. The location of the maximum of the first peak and its intensity are similar for all samples studied and only some changes in the width of this peak are observed [169].

2.3.16   COMPARISON BETWEEN DIFFERENT MATHEMATICAL MODELS

This chapter has reviewed modeling over the years and the progress that can be identified. It has been demonstrated that modelers, using modern computational systems, base their calculations on structure systems, usually around the graphitic micro crystallite. An objective of such modelers is the simulation of adsorption isotherms determined experimentally. Whether or not the structural models assumed for their work can be considered to be "realistic" is rarely a matter for discussion. Accordingly, the considerations of such modelers should not be adopted, uncritically, by those who have other interests in activated carbon. There is no unique structure within an activated carbon, which provides a specific isotherm, for example the adsorption of benzene at 273K. The isotherm is a description of the distribution of adsorption potentials throughout the carbon, this distribution following a normal or Gaussian distribution. If a structure is therefore devised which permits a continuous distribution of adsorption potentials, and this model predicts an experimental adsorption isotherm, this then is no guarantee that the structure of the model is correct. The wider experience of the carbon scientist, who relates the model to preparation methods and physical and chemical properties of the carbon, has to pronounce on the "reality" or acceptance value of the model. Unfortunately, the modeler appears not to consult the carbon chemist too much, and it is left to the carbon chemist to explain the limited acceptability of the adopted structures of the modeler. This approach relies on different assumptions in order to obtain relationships allowing the calculation of the main characteristics of adsorbent structural heterogeneity. The assumptions are summarized as follows:

(a) The shape of the pores is assumed to be slit like with effects of inter connectivity to be neglected.

(b) The molecule-surface interactions obey the equation of steel.

(c) The adsorbed phase can be considered as liquid-like layers between the two parallel walls of the pore, as a monolayer or as a double layer.

(d) The thickness of the adsorbed layer is determined by the distance between two parallel walls of the slit-like porosity, the density of the layers being constant between the layers and being equal to that of the bulk liquid.

(e) No gas-phase adsorbate exists within the volumes between the layers [1, 2, 133].

Earlier, researchers used GCMC and DFT as modeling methods based on slit-like pores with graphitic surfaces, concluded that the inclusion of surface heterogeneity into the model would have little effect on obtained simulations. Further, it is considered that the use of alternative geometries may be worth pursuing. Several quite different approaches have been brought to the modeling of activated carbons. The concept of porosity not accessible to nitrogen at 77 K is understood. A different concept, that the surface structure (in terms of pore density) is not representative of the bulk density of porosity, can be commented upon adversely. The suggestion that the fibrils of botanical structure of the olive stones have a role to play in maintaining the mechanical strength of the stones may need to be modified in the light of modem analyzes. The approach of this Section is limited because of confusion over what acceptable surface area values are and what not acceptable surface area values are. In addition, this section makes two assumptions which can be commented upon critically, namely that carbons can be categorized into well-defined groups according to their pore volumes and mean pore dimensions, and that it is adsorption on graphitic micro crystallites which controls the creation of an isotherm. Finally, the use of two mazes, two-dimensional and three-dimensional, designed as puzzle games provides helpful models to understand the complexity of the network of micro porosity within a carbon. The simple ratio of carbon atoms to nitrogen molecules adsorbed as a monolayer capacity could be useful in assessments of how the micro porosity fits into the carbon layer networks. This PSD calculating Models designed, to describe the micro porous nature of carbons have been compared and contrasted leading to an assessment of the requirements of a comprehensive model to account for the properties of micro porous systems. No comprehensive model has, as yet, been created. These of a model based on a maze, provides insights. In the different adsorption processes, both in gas and liquid phase, the molecules or atoms absorbable) are fixed (adsorbed) on the carbon (adsorbent) surface by physical interactions (electrostatic and dispersive forces) and/ or chemical bonds. Therefore, a relatively large specific surface area is one of the most important properties that characterize carbon adsorbents. The surface of the activated carbons consists mainly of basal planes and the edges of the planes that form the edges of micro crystallites. Adsorption capacity related parameters are usually determined from gas adsorption measurements. The specific surface area is calculated by applying the BET equation to the isotherms generated during the adsorption process. The adsorption of N_2 at 77 K or CO_2 at 273K are the most commonly used to produce these isotherms. The BET theory is based upon the assumption that the monolayer is located on surface sites of uniform adsorption energy and multilayer build-up via a process analogous the condensation of the liquid adsorbate. For convenience, the BET equation is normally expressed in the form which requires a linear relationship between p/p_0, and model parameters from which the monolayer capacity, nm (mol g^{-1}), can be calculated. In activated carbons the range of linearity of the BET plot is severely restricted to the p/p_0 range of 0.05–0.20. The alternative form of linearization of the BET equation appears to be more convenient for a micro porous solid since the choice of the appropriate experimental interval is free of ambiguity. The BET equation, however, is subject to various limitations when applied to micro porous carbons. Thus, constrictions in the micro porous network may cause molecular sieve effects and molecular shape selectivity [1, 2, 133].

Diffusion effects may also occur when using N_2 at 77 K as adsorbate since at such low temperatures the kinetic energy may be in sufficient to penetrate all the micro pores. For this reason adsorption of CO_2 at higher temperatures (273K) is also used. CO_2 and N_2 isotherms are complementary. Thus, whereas from the CO_2 isotherm micro pores of up to approximately 10^{-9} m width can be measured, the N_2 can be used to test larger pores. Despite these limitations the BET surface area is the parameter most commonly used to characterize the specific surface area of carbon adsorbents. On the basis of volume-filling mechanism and thermodynamic considerations, Dubinin and Radushkevich found empirically that the characteristic curves obtained using the Potential Theory for adsorption on many micro porous carbons could be linear zed using the DR equation. For some micro porous carbons the DR equation is linear over many orders of magnitude of pressure. For others, however, deviations from the DR equation are found. For such cases the Dubinin-Astakhov equation has been proposed in which the exponent of the DR equation is replaced by a third adjustable parameter, n, where $1 < n < 3$. Both the BET and the Dubinin models are widely thought to adequately describe the physical adsorption of gases on solid carbons. BET surface areas from many micro porous carbons range from 500 to 1500 $m^2 g^{-1}$. However, values of up to 4000 $m^2 g^{-1}$ are found for some super activated carbons and these are unrealistically high. The relatively high values of the surface areas of activated carbons are mainly due to the contribution of the micro pores and most of the adsorption takes place in these pores. At least 90–95% of the total surface area of an activated carbon may correspond to micro pores. However, me so and macro pores also play a very important role in any adsorption process since they serve as the passage through which adsorbate reaches the micro pores. Thus, mesopore, which branch off from the macro pores, serve as passages for the adsorptive to reach the micro pores. In such mesopore capillary condensation may occur with the formation of a meniscus in adsorbate. Although the surface area of the mesopore is relatively low inmost activated carbons, some may have well-developed mesoporosity (200 m² g⁻¹ or even more). In addition, depending on the size of the adsorbate molecules, especially in the case of some organic molecules of a large size, molecular sieve effects may occur either because the pore width is narrower than the molecules of the adsorbate or because the shape of the pores does not allow the molecules of the adsorbate to penetrate into the micro pores. Thus, slit-shaped micro pores formed by the spaces between the carbon layer planes are not accessible to molecules of a spherical geometry, which have a diameter larger than the pore width. This means that the specific surface area of a carbon is not necessarily proportional to the adsorption capacity of the activated carbon. Pore size distribution, therefore, is a factor that cannot be ignored. The suitability of a given activated carbon for a given application depends on the proportion of pores of a particular size. In general highly micro porous carbons are preferred for the adsorption of gases and vapors and for the separation of gas molecules of different dimensions if the carbon possesses a suitable distribution of narrow size pores (molecular sieves)while well developed me so and macro porosity is necessary for the adsorption of solutes from solutions [145, 161].

It is the entrance dimension and shape, which controls the adsorption process, be hexagonal (an appropriate shape), or circular or square. Once the adsorbate molecule is through the pore entrance, then the characteristics of the adsorbent take over and the isotherm is created that is not enough, because the processes of physical and chemical activation have to be understood and to do this, requires three-dimensional models. Porosity in carbons it would be an advantage to have some idea of the structure of this network include of carbon atoms, in three dimensions, in order to understand the extraction (gasification) process. There are four limitations of importance. First, this maze, of course, is in two dimensions; second, the lines of the maze are too orientated relative to an x–y axis. Such parallelism is unlikely to exist within a porous carbon; third, this labyrinth is best suited to a micro porous carbon, only and not to micro porous carbon fibers. Fourth, in such a model, rates of diffusion are likely to be too slow and hence there is a need to consider the location of mesoporosity. The inclusion of mesoporosity is another matter. Mesoporosity has to promote enhanced adsorption to the interior of the fiber. As a matter of scaling, although the models that mentioned before provide an impressive number of adsorption locations, it will require too much of such models. The human mind cannot cope with this necessity. But, apart from these limitations the similarities are relevant enough:

1. There is a continuous connection between the lines throughout the labyrinth. All carbon atoms form part of a continuous graphene sheet.
2. There is a continuous connection of the routes (spaces) of the labyrinth. Hence, all adsorption sites are available to adsorbate molecules.
3. The widths of the routes (spaces) of the labyrinth are not constant. Some are narrower than other. This is a very relevant point as it demonstrates, very clearly, the range of porosities, within the definition of micro porosity of <2.0 nm, this accounting for molecular sieving effects.
4. Some of the routes of the labyrinth are barely visible (being very close to each other). This offers the suggestion that it is representative of closed porosity, that is closed to everything except helium and lithium, noting, on the way, that the term closed porosity is an imprecise term, meaning porosity not accessible to a specifically defined adsorbate molecule.
5. The PSD calculating models makes the point that access to the interior of the labyrinth is available from all external surfaces.
6. A close inspection of the edges of the lines of labyrinth (surfaces of the carbon surfaces) indicates a lack of smoothness, the edges being rough. This point is of importance to carbon science because the surfaces of porosity are of imperfect graphene-like layers, with surface irregularities and other defects.
7. This models that mentioned before, also demonstrates that most of the continuous three-dimensional graphene layer is able to act as an adsorbent surface making use of both sides. Rarely, do parts of the graphene layer come together (stack) in two or three layers. This requirement other the range of such "different" sites is a feature of the models. Clearly, all the requirements for molecular-sieve properties, the dynamics and enthalpies of adsorption are present even

though the detail is absent. The effect of increasing HTT would be to remove defects or irregularities of the carbon atom network resulting in decreased enthalpies of adsorption on more homogeneous surfaces. Such a porous system would respond to SAXS and SANS with the range of ring structures in the carbon atom network being sufficient to explain observed Raman spectra and electron spin resonance (ESR) during the carbonization process [1, 133].

There are many methods for calculation of PSDs and most of them are potentially applicable for nano porous carbons. PSDs for nano porous carbons are usually evaluated using methods based on either the Kelvin equation or the Horvath-Kawazoe method and its modifications. The first group includes the models of Barrett, Joyner and Halenda (BJH) method, Cranston and Inkley (CI), Dollimore and Heal (DH), and Broekhoff and de Boer (BdB). Although the BJH, CI, and DH methods are often considered as appreciably different, all of them are based on the general concept of the algorithm outlined in the BJH work. To implement the algorithms proposed in these three methods, the knowledge of the relation between the pore size and capillary condensation or evaporation pressure and the t-curve is required, and a choice needs to be made which branch of the isotherm is appropriate for the PSD calculation. The original BJH, CI, and DH models are not fully consistent as far as the selection of these relations and the choice of the branch of the isotherm are concerned. These inconsistencies are capable of affecting the results of calculations much more than the minor differences in the algorithms, being most likely responsible for claims that these three methods differ substantially. The BJH, CI and DH methods assume the same general picture of the adsorption desorption process. Adsorption in meso pores of a given size is pictured as the multi layer adsorption followed by capillary condensation (filling of the pore core, that is, the space that is unoccupied by the multilayer film on the pore walls) at a relative pressure determined by the pore diameter. The desorption pictured as capillary evaporation (emptying of the pore core with retention of the multilayer film) at a relative pressure related to the pore diameter followed by thinning of the multi layer. Because the concept underlying the BJH, CI, and DH models appears to be correct, it is important to:

1. establish an accurate relation between the pore size and capillary condensation or evaporation pressure;
2. determine to correct t-curve;

3. verify whether adsorption or desorption, or both branches of the isotherms, are suitable for the accurate pore size assessment.

This would allow performing accurate PSD calculations using these simple algorithms. Theoretical considerations, NLDFT calculations, computer simulations, and studies of the model adsorbents strongly suggested that the Kelvin equation commonly used to provide a relation between the capillary condensation or evaporation pressure and the pore size calculating. Porosity of nano porous carbonaceous materials is usually analyzed on the basis of nitrogen adsorption isotherms, which reflect the gradual formation of a multilayer film on the pore walls followed by capillary condensation in the unfilled pore interior. The pressure dependence of the film thickness is affected by the adsorbent surface. Hence, an accurate estimation of the pore size distribution requires a correction for the thickness of the film formed on the pore walls. The latter (so-called t-curve) is determined on the basis of adsorption isotherms on nonporous or macro porous adsorbents of the surface properties analogous to those for the adsorbent studied. Modified nonlocal density functional theory (MDFT) has been shown to provide an excellent description of the physical adsorption of nitrogen or argon on the energetically uniform surface of graphite. This and other formulations of DFT have been used to model adsorption in narrow slit pores and provide the basis for a method of estimating pore size distribution from experimental isotherms [1, 2, 145, 161].

Important parameters that greatly affect the adsorption performance of a porous carbonaceous adsorbent are porosity and pores structure. Consequently, the determination of PSD of coal-based adsorbents is of particular interest. For this purpose, various methods have been proposed to study the structure of porous adsorbents. A direct but cumbersome experimental technique for the determination of PSD is to measure the saturated amount of adsorbed probe molecules, which have different dimensions. However, there is uncertainty about this method because of networking effects of some adsorbents including activated carbons. Other experimental techniques that usually implement for characterizing the pore structure of porous materials are mercury porosimetry, X-ray diffraction (XRD) or small angle X-ray scattering (SAXS), and immersion calorimetry. A large number of simple and sophisticated models have been presented to obtain a really estimation of PSD of porous adsorbents. Relatively simple but restricted applicable models such as BJH, DH, Mikhail et al. (MP), HK, JC, Wojsz and Rozwadowski (WR), Kruk-Jaroniec-Sayari (KJS), and ND were presented from 1951 to 1999 by various researchers for the prediction of PSD from the adsorption isotherms. For example, the BJH method, which is usually recommended for meso porous materials is in error even in large pores with dimension of 20 nm. The main criticism of the MP method, in addition to the uncertainty regarding the multilayer adsorption mechanism in micro pores, is that we should have a judicious choice of the reference isotherm. HK model was developed for calculating micro pore size distribution of slit-shaped pore; however, the HK method suffers from the idealization of the micro pores filling process. Extension of this theory for cylindrical and spherical pores was made by Saito and Foley and Cheng and Ralph. By applying some modifications on the HK theory, some improved models for calculating PSD of porous adsorbents have been recently extended the Nguyen and Do method for the determination of the bimodal PSD of various carbonaceous materials from a variety of synthetic and experimental data. The pore range of applicability of this model besides other limitations of ND method is its main constraint [1, 2, 133, 162].

Many researchers determined the PSD of activated carbons based on liquid chromatography (LC). Choice of suitable solvent and pore range of applicability of this method are two main problems that restrict its general applicability. More sophisticated methods such as molecular dynamics (MD), Monte Carlo simulation, GCMS, and density functional theory (DFT) are theoretically capable of describing adsorption in the pore system. The advantages of these methods are that they can apply on wide range of pores width. But, they are relatively complicated and provide accurate PSD estimation based on just some adsorbents with specified shapes. Requiring extensive computation time and significantly different idealized conditions from a real situation are other drawbacks of such methods. Other researchers used a new approach based on Monte Carlo integration to derive pore size and its volume distribution for porous solids having known configuration of solid atoms. They applied the proposed method to a wide range of commonly used porous solids. However, this method also seems to have limitations of Monte Carlo simulation in addition to requiring the information of solid atomistic configuration. Recently, some researchers proposed a multi scale approach based on GCMC to predict the adsorption isotherms and PSD of porous solids. However, the proposed methodology needs further improvement to provide high accuracy predictions besides mass of computations that is needed. Kind of PSD assumes in DA equation itself, is important in design of PSD modeling. Does this distribution change if the parameters of the DA isotherm change (especially n). Some researchers studied the simulated nitrogen adsorption isotherms in heterogeneous carbons and assumed Gaussian distribution of pores. It was shown that the simulated isotherms can be fitted by typical DR, although not in the whole range of relative pressures. Two groups of low-temperature (= 77.5 K) N_2 adsorption isotherms were generated, and studied the influence of n at constant E_0 and, to the contrary, the effect of E_0 at constant n. The obtained curves were converted into high-resolution -plots in order to explain the mechanism of adsorption. Moreover, the new algorithm (called the adsorption stochastic algorithm (ASA) was used to solve the problem of fitting the local adsorption isotherms of the Nguyen and Do method to experimental data. The obtained results show that the DA equation generates isotherms describing almost a homogeneous structure of pores and/or a bimodal heterogeneous structure. Corresponding PSDs indicate the presence of homogeneous porosity, primary and/or secondary micro pores filling, or both. The parameter of the DA equation is responsible not only for the homogeneity of pores (the deviation of pores from average size) but for the adsorption mechanism in micro pores. In other words, lowering n leads to the change in this mechanism from primary to simultaneous primary and secondary micro pores filling. Taking all obtained results into account suggests that the DA equation is probably the most universal description of adsorption in micro pores. All theoretical models (e.g., as considered in the current study: HK, ND, and DFT or etc.) are connected with their own specific assumptions of the description of the porous structure and/or mechanism of adsorption. Of course, those postulations can significantly influence obtained results, PSD curves. The modeling of local and/or global adsorption isotherms for different pore widths using DFT theory, Monte Carlo simulations, and the ND model leads to results that should be, for some cases, treated with caution. The main simplifications of those theories are, for example, the neglect of the pore connectivity, ignoring of different thickness of carbon micro crystallites forming micro pores, or existence of various surface groups on the surface of activated carbons. Moreover, it is very difficult to find papers where authors obtained satisfactory results (using the abovementioned theories) describing simultaneously the experimental adsorption isotherm, adsorption enthalpy, and entropy (or heat capacity) for adsorption in micro porous carbons around room temperature. Although all simulation and modeling of carbons is very interesting and sometimes spectacular, it should be remembered that small changes in the values of fundamental parameters taken as constants in calculations can lead to drastic changes in the results obtained. In this review, the results of this type of the calculation are speculative as long as a satisfactory model of the structure of carbons is not evaluated. It should be pointed out that very complicated models of the structure of a micro porous activated carbon are sometimes considered in some advanced numerical and simulation calculations (e.g., the RMC method, where the surface sites have been added at random points on the edges of the graphene micro crystals characterizing by differing size and shape) in order to describe the "real" structure of activated carbons. However, taking into account this complex structure is connected with considerable extension of the time of calculations. On the other hand, very puzzling results are obtained from the description of the micro pore structure of various carbonaceous materials (the different origin and thermal (or chemical) treatment)) for the reason that the differences between the micro pore size distribution plots are insignificant. For example, on the results of micro porosity determination from DFT method for three carbons, the number of peaks on PSD curves and the ranges of their location, as well as the shapes, are very similar. In our opinion this behavior of PSDs is very surprising and can be caused by the low sensitivity and the simplifications of the mentioned above methods [145, 161].

The advantage of these models is that they are experimentally convenient and do not need complicated PSD calculations. In 1998, Bhatia successfully applied the combination of finite element collocation technique with regularization method to extract various double peak PSDs from synthetic isotherm data points contaminated with 1% normally distributed random errors using DR isotherm. They applied the constraint of non-negativity of solutions by simply using a Newton-Rap son technique. Although they reported that the method is stable over a wide range of values of the regularization parameter, the application of non-negativity constraint usually provides unrealistic solutions. Some researchers proposed a new method based on the modification of DR equation by introducing adsorption density and correlating between the pores filling pressure and critical pore size for nitrogen adsorption at 77 K. The results are found comparable with other popular PSD methods such as MP, JC, HK, and DFT. In addition to uncertainty about the general performance of this model as a result of some assumptions in the model derivation, it is relatively complicated and the procedure for obtaining PSD is cumbersome. The average diameter of the mesopore is usually calculated from the nitrogen adsorption data using Kelvin equation. Recently, researchers proposed two new algorithms (SHN1, SHN2) for reliable extraction of PSD from adsorption and condensation branch of isotherms. According to their results, it seems that these two methods have also some limitation similar to the other previous ones. The regularization technique that used for obtaining optimum value of regularization parameter is challengeable in these models. In addition, the basis of these methods is Kelvin equation, which unlikely provides reliable PSD for micro porous solids. Although much has been done to address the PSD of porous adsorbents, up to now, no general reliable theory is available leading to the conclusion that for micro porous carbons the extensive investigation should still continue. In the recent study works, have tried as a novel work to extend the analysis on PSD of porous carbon nano adsorbents by investigating the effects of different parameters on it. Some well-known models based on Dubinin's method namely, DS and Stoeckli models, etc., were used to investigate the effects of these parameters on the porosity of AC samples, and the results were compared with the two widely used methods of HK and improved IHK for the determination of PSD of micro or mesoporous solids. These models were derived based on benzene as a reference adsorbate, or N_2 or Ar, because this gases provides more accurate estimations than other adsorbents. On this basis, adsorption isotherm data of benzene at 30 °C or N_2 in 77 K or Ar in 87 K were used to determine the PSD of each carbon samples. Table 2.1 summarized novel models for characterization of pores and determines the PSD in carbon materials [1, 2, 133, 162].

BET and the Dubinin models (DRS, DA) beside HK and DFT models are widely thought to adequately describe the adsorption process of carbon nano adsorbents. For convenience, the BET equation is normally expressed in the form, which requires a linear relationship between p/p_0, and model parameters from which the monolayer capacity, and adsorption parameters can be calculated, in addition this method is more accessible. However, by compare of different model and simulation methods, this model is subject to various limitations applied to micro porous carbons to calculating PSD in CNT-Textile composite.

KEYWORDS

- **Adsorption stochastic algorithm**
- **Carbon Nano Adsorbents**
- **Mesopore**
- **Micro crystals characterizing**
- **Model parameters**
- **Monolayer capacity**

CHAPTER 3

SIMULATION

CONTENTS

3.1   INTRODUCTION

Simulation is the imitation of the operation of a real world or system over time to develop a set of assumptions of mathematical, logical and symbolic relationship between the entities of interest of the system, to estimate the measures of performance of the system with the simulation generated data. Simulation can be used to verify analytic solutions. By simulating different capabilities for a machine requirements can be determined. Simulation is not appropriate in some cases include:

- when the problem can be solved using common sense and the problem can be solved analytically;
- when it is easier to perform direct experiments and the simulation costs exceed the savings;
- when the resources or time are not available and system behavior is too complex or can't be defined;
- when there isn't the ability to verify and validate the model.

Simulation is the appropriate tool in some cases include:

- simulation can be used to experiment with new designs or policies prior to implementation, so as to prepare for what may happen;
- simulation can be used to verify analytic solutions and by simulating different capabilities for a machine, requirements can be determined.

Advantages of simulation include:

- new polices, operating procedures, decision rules, information flows, organizational procedures, and soon can be explored without disrupting ongoing operations of the real system;
- new hardware designs, physical layouts, transportation systems, and soon, can be tested without committing resources for their acquisition;
- hypotheses about how or why certain phenomena occur can be tested for feasibility;
- insight can be obtained about the interaction of variables;
- insight can be obtained about the importance of variables to the performance of the system;
- a simulation study can help in understanding how the system operates rather than how individuals think the system operates;
- "what-if "questions can be answered. This is particularly useful in the design of new system;

- the two main families of simulation technique are molecular dynamics (MD) and Monte Carlo (MC); additionally, there is a whole range of hybrid techniques which combine features from both [161–165].

For example, in recent study works about PSD calculating simulated samples, the average of numbers adsorbed molecules fluctuates during the simulation calculating of for a range of chemical potentials enables the adsorption isotherm to be constructed. The walls of the slit pores lie in the x–y plane. Normal periodic boundary conditions, together with the minimum image convention, are applied in these two directions. For low pressures, P/P_0 that smaller than 0.02, the length of the simulation cell in the two directions parallel to the walls was maintained at 10 nm for each of the pore widths studied, to maintain a sufficient number of adsorbed molecules. For higher pressures where more water molecules were present, the minimum cell length in the x and y directions was 4 nm. The average number of water molecules in the simulation cell varied from a few molecules at the lowest pressures to a few hundred or thousands of molecules when the pores were full; filled pores contained about 320 molecules for a pore width of 0.79 nm, 460 at 0.99 nm, 830 at 1.69 nm) Calculations were carried out on the Cornell Theory Center IBM SP2. In determining the adsorption isotherm, commenced with the cell empty; a value of the fugacity corresponding to a low pressure was chosen and the average adsorption determined from the simulation. The final configuration generated at each stage was used as the starting point for simulations at higher fugacities. The pressure of the bulk gas corresponding to a given chemical potential was determined from the ideal gas equation of state. Gas phase densities corresponding to the range of chemical potentials studied were determined by carrying out simulations of the bulk gas. These were found to agree with those calculated from the ideal gas equation within the estimated errors of the simulations [161–165].

3.2 MONTE CARLO SIMULATION

In a Monte Carlo simulation we attempt to follow the 'time dependence' of a model for which change, or growth, does not proceed in some rigorously pre-denned fashion (according to Newton's equations of motion) but rather in a stochastic manner which depends on a sequence of random numbers which is generated during the simulation method, in a simulation cell that design for special purposes. With a second, different sequence of random numbers the simulation will not give identical results but will yield values, which agree with those obtained from the first sequence to within some 'statistical error.' A very large number of different problems fall into this category: in percolation an empty lattice is gradually filled with particles by placing a particle on the lattice randomly with each 'tick of the clock. Considering problems of statistical mechanics, we may be attempting to sample a region of phase space in order to estimate certain properties of the model, although we may not be moving in phase space along the same path, which an exact solution to the time dependence of the model would yield. Remember that the task of equilibrium statistical mechanics is to calculate thermal averages of (interacting) many-particle systems: Monte Carlo simulations can do that, taking proper account of statistical errors and their effects in such systems. Many of these models will be discussed in more detail in later chapters so we shall not provide further details here. Since the accuracy of a Monte Carlo estimate depends upon the thoroughness with which phase space is probed, improvement may be obtained by simply running the calculation a little longer to increase the number of samples. Unlike in the application of many analytic techniques (perturbation theory for which the extension to higher order may be prohibitively difficult), the improvement of the accuracy of Monte Carlo results is possible not just in principle but also in practice [150, 156, 163, 165].

The method of Monte Carlo simulation has proved very useful for studying the thermodynamic properties of model systems with moderately many degrees of freedom. The idea is to sample the system's phase space stochastically, using a computer to generate a series of random configurations. We take the phase space to consist of N discrete states (with label i), though the method applies equally to continuous systems. Often only a tiny fraction of the phase space (the part at low energy) is relevant to the properties being studied, due to the strong variation of the Boltzmann equation) in the canonical ensemble (CE). It is then helpful to sample in an ensemble (with relative weights w_i and absolute probabilities $p_i = w_i / P_j w_j$), which is concentrated on this region of phase space. The Metropolis algorithm samples directly in the CE, and is good at determining many physical properties [12, 17, 29].

The price to be paid for this is that successive configurations are not independent (typically they have a single microscopic difference), but instead form a Markov chain with some equilibration time teq (w_i). We may distinguish two important characteristics of a Monte Carlo simulation: its ergodicity (measured by teq (w_i)) and its pertinence (measured by Ns $(w_i;$ I), the average number of independent samples needed to obtain the information I that we seek). We should choose w_i so as to minimize the total number of configurations that need to be generated, which is proportional to teq (w_i) Ns $(w_i; I)$. It is easy to specify an ensemble, which would yield the sought information if independent samples could be drawn from it, but an ensemble with too much weight at low energies may become fragmented into "pools" at the bottoms of "valleys" of the energy function, and so have a large equilibration time. For example, it is well-known that at low temperatures the Metropolis algorithm can get stuck in ordered or glassy phases. Ergodicity may be improved by sampling instead in a nonphysical ensemble with a broad energy distribution, which allows the valleys to be connected by paths passing through higher energies. A weight assignment leading to such a distribution cannot in general be written as an explicit function of energy alone; rather it is an algorithm's purpose to find this assignment, which then tells us about the density of states (E). This reversal (starting with the distribution and finding the weights) of the usual Monte Carlo process can be achieved using a series of normal simulations, adjusting the weight w_i after each run so that the resulting energy distribution w_i (E) converges to the desired one. The last application reported here is a simulation of a regular system with frustration, the triangular antiferromagnetic, on a 48×48 parallelogram with periodic boundary conditions. Using 5 runs of 7.4×105 sweeps, we obtained a ground state entropy of 0.32320, with a variance of 0.00015, which is consistent with the exact bulk value$\simeq$0.32307 [163–165]. As computers have improved in capability, the simulation of large statistical systems governed by known Hamiltonians has become an important tool of the theoretical physicist. Applications range from studies of phase transitions in condensed matter to calculation of hadronic properties via lattice gauge theory. Most of these simulations rely on adaptations of the algorithm of Metropolis. This generates a sequence of configurations via a Markov process such that ultimately the probability of encountering any given configuration in the sequence is proportional to the Boltzmann weight functions (C), where is the energy for a statistical mechanics problem or the action for a quantum field theory simulation. Thus one obtains a sample of configurations, which dominate the partition function sum or path integral. An alternative simulation technique is the molecular dynamics or micromechanical method. This begins with a set of equations for a dynamical evolution, which conserves the total energy. Upon numerical integration the system will flow through phase space in a hopefully ergodic manner(indeed, nonergodic behavior would represent a fascinating exception to the generic case). Such a program does not explicitly depend on an inverse temperature, which is determined dynamically by measuring, say, the average kinetic energy, which by equipment partition should be IkT per degree of freedom. Note that the conventional micromechanical simulations make no use of random numbers, which are effectively generated by the complexity of the system. This technique has recently been applied to lattice gauge theory. Such equation gives another micro canonical formulation, which was discussed in the context of continuum field theory by Strominger [168].

Monte Carlo methods are used as computational tools in many areas of chemical physics. Although this technique has been largely associated with obtaining static, or equilibrium properties of model systems, Monte Carlo methods may also be used to study dynamical phenomena. Often, the dynamics and cooperatively leading to certain structural or configurationally properties of matter are not completely amenable to a macroscopic continuum description. On the other hand, molecular dynamics simulations describing the trajectories of individual atoms or molecules on potential energy hyper surfaces are not computationally capable of probing large systems of interacting particles at long times. Thus, in a dynamical capacity, Monte Carlo methods are capable of bridging the ostensibly large gap existing between these two well-established dynamical approaches in set he "dynamics" of individual atoms and molecules are modeled in this technique, but only in coarse-grained way representing average features which would arise from a lower-level result. The application of the Monte Carlo method to the study of dynamical phenomena requires a self-consistent dynamical interpretation of the technique and a set of criteria under which this interpretation may be practically extended. In recent publications, certain inconsistencies have been identified which arise when the dynamical interpretation of the Monte Carlo method is loosely applied. These studies have emphasized that, unlike static properties, which must be identical for systems having identical model Hamiltonians, dynamical properties are sensitive to the manner in which the time series of events characterizing the evolution of a system is constructed. In particular, Monte Carlo studies comparing dynamical properties simulated away from thermal equilibrium have revealed differences among various sampling algorithms. These studies have underscored the importance of using a Monte Carlo sampling procedure in which transition probabilities are based on a reasonable dynamical model of a particular physical phenomenon under consideration, in addition to satisfying the usual criteria for thermal equilibrium. Unless transition probabilities can be formulated in this way, a relationship between Monte Carlo time and real time cannot be clearly demonstrated. In many Monte Carlo studies of time-dependent phenomena, results are reported in terms of integral Monte Carlo steps, which obfuscate a definitive role of time. Ambiguities surrounding the relationship of Monte Carlo time to real time preclude rigorous comparison of simulated results to theory and experiment, needlessly restricting the technique. Within the past few years, the idea that Monte Carlo methods can be used to simulate the Poisson process has been advanced in a few publications and some Monte Carlo algorithms, which are implicitly based on this assumption have been used. An attractive prospect, since within the theory of Poisson processes the relationship between Monte Carlo time and real time can be clearly established [173].

Early use of Monte Carlo techniques was made for the quantitative evaluation of fault trees. While some effort has continued in the use of purely Monte Carlo methods, they have largely been supplanted deterministic techniques often referred to as Kinetic Tree methods. Two limitations, however, present themselves in the use of Kinetic Tree methods. First, in Kinetic Tree methods the reliability characteristics of each component are modeled separately. To evaluate the fault tree by combining component failure probabilities, the components are assumed to behave independently of one another. In fact, dependencies often arise from common mode failures, from the increased stress in partially disabled systems, and from a variety of errors in testing, maintenance and repair. Due to this limitation of the Kinetic Tree formulation there is increasing use of Markov models for reliability analysis, for with such models quite general dependencies between components may be treated. For systems with more than a few components, however, Markov analysis by deterministic means becomes a prodigious task. For even while innovative methods have been employed to reduce the complexity of the computations, the fact remains that one must solve a set of 21 coupled first-order differential equations, thus even a system with only ten components will result in a system of over one thousand coupled equations with a transition matrix with over a million elements. Moreover, if some of the components are repairable, the equations are likely to be quite stiff, requiring that very small time steps be used in the numerical integration. A second limitation on Kinetic Tree methods is a result of the lack of precision to which the component failure and repair rates are normally known. Invariably this is accomplished by Monte Carlo sampling of the failure rate data using log-normal or other distributions. The fault tree is evaluated deterministically with data from each data sampling, and the mean, variance and other characteristics of the system are estimated. A similar procedure is also applied to Markov models, requiring that the solution of the coupled set of differential equations be repeated thousands of times. What follows is the formulation of a class of Monte Carlo methods, which provides a natural framework for the treatment of both component dependencies and data uncertainties. Some researchers formulate Monte Carlo simulation of the unreliability of systems with repairable components within the framework of a Markov process. This approach retains the power of deterministic Markov methods in modeling component dependencies that would not be possible if direct Monte Carlo simulation were to be carried out. At the same time the Monte Carlo simulation requires very little computer memory. Variance reduction techniques, similar to those that have been highly developed for neutral particle transport calculations, are applied to greatly increase the computational efficiency of Monte Carlo reliability calculations. Monte Carlo formulation is generalized to include probability distributions that represent the uncertainty in the component failure and repair rate data. The variance in the result is then due to two causes: the finite number of random walk simulations, and the uncertainty in the data. A batching technique is introduced and is shown to further reduce that part of the variance due to the finite number of random walks without a commensurate increase in computing effort. The Markov Monte Carlo formulation was extended to problems with some data uncertainties, and a batching technique was shown to lead to further improvements in the figure of merit. While we have not had the opportunity to make numerical comparisons between Markov Monte Carlo and Kinetic Tree methods, which use Monte Carlo data sampling, an observation seems in order. For equal data sampling one would equate the number of Markov Monte Carlo batches to the number of Kinetic Tree trials. If one then chose the batch size just large enough so that the random walk variance could be ignored relative to the variance due to data uncertainty fair comparison of computational efficiency would be the Monte Carlo time per batch versus the Kinetic Tree time per trial.

This, of course, assumes that the problem is chosen in which component dependencies do not rule out the use of Kinetic Tree methods [174].

Within the contents of this book we have attempted to elucidate the essential features of Monte Carlo simulations and their application to problems in statistical physics. We have attempted to give the reader practical advice as well as to present theoretically based background for the methodology of the simulations as well as the tools of analysis. New Monte Carlo methods will be devised and will be used with more powerful computers, but we believe that the advice given to the reader that will remain valid. In general terms we can expect that progress in Monte Carlo studies in the future will take place along two different routes. First, there will be a continued advancement towards ultra high-resolution studies of relatively simple models in which critical temperatures and exponents, phase boundaries, etc., will be examined with increasing precision and accuracy. As a consequence, high numerical resolution as well as the physical interpretation of simulation results may well provide hints to the theorist who is interested in analytic investigation. On the other hand, we expect that there will be a tendency to increase the examination of much more complicated models, which provide a better approximation to physical materials. As the general area of materials science blossoms, we anticipate that Monte Carlo methods will be used to probe the often-complex behavior of real materials. This is a challenge indeed, since there are usually phenomena, which are occurring at different length and time scales. As a result, it will not be surprising if multi scale methods are developed and Monte Carlo methods will be used within multiple regions of length and time scales. We encourage the reader to think of new problems which are amenable to Monte Carlo simulation but which have not yet been approached with this method. Lastly, it is likely that an enhanced understanding of the significance of numerical results can be obtained using techniques of scientific visualization. The general trend in Monte Carlo simulations is to ever-larger systems studied for longer and longer times. The mere interpretation of the data is becoming a problem of increasing magnitude, and visual techniques for probing the system (again over different scales of time and length) must be developed. Coarse-graining techniques can be used to clarify features of the results, which are not immediately obvious from inspection of columns of numbers. 'Windows' of various size can be used to scan the system looking for patterns, which develop in both space and time; and the development of such methods may well profit from interaction with computer science. Clearly improved computer performance is moving swiftly in the direction of parallel computing. Because of the inherent complexity of message passing, it is likely that we shall see the development of hybrid computers in which large arrays of symmetric (shared memory) multiprocessors appear (Until much higher speeds are achieved on the Internet, it is unlikely that nonlocal assemblies of machines will prove useful for the majority of Monte Carlo simulations.). We must continue to examine the algorithms and codes, which are used for Monte Carlo simulations to insure that they remain well suited to the available computational resources. We strongly believe that the utility of Monte Carlo simulations will continue to grow quite rapidly, but the directions may not always be predictable. We hope that the material in this book will prove useful to the reader who wanders into unfamiliar scientific territory and must be able to create new tools instead of merely copying those that can be found in many places in the literature [165].

3.2.1   RANGE OF PROBLEMS CAN BE SOLVED USING MONTE CARLO SIMULATION

The range of different physical phenomena, which can be explored using Monte Carlo methods, is exceedingly broad. Models which either naturally or through approximation can be discretized can be considered. The motion of individual atoms may be examined directly, for example, in a binary (AB) metallic alloy where one is interested in inter diffusion or un mixing kinetics (if the alloy was prepared in a thermodynamically unstable state) the random hopping of atoms to neighboring sites can be modeled directly. This problem is complicated because the jump rates of the different atoms depend on the locally differing environment. Equilibrium properties of systems of interacting atoms have been extensively studied as have a wide range of models for simple and complex fluids, magnetic materials, metallic alloys, adsorbed surface layers, etc. More recently polymer models have been studied with increasing frequency; note that the simplest model of a flexible polymer is a random walk, an object which is well suited for Monte Carlo simulation. Furthermore, some of the most significant advances in under-standing the theory of elementary particles have been made using Monte Carlo simulations of lattice gauge models [131, 163–165].

3.2.2   THE PROBLEMS OF MONTE CARLO SIMULATION

3.2.2.1   LIMITED COMPUTER TIME AND MEMORY

Because of limits on computer speed there are some problems, which are inherently not suited to computer simulation, at this time. A simulation, which requires years of CPU time on whatever machine is available, is simply impractical. Similarly a calculation, which requires memory, which far exceeds, that which is available can be carried out only by using very sophisticated programming techniques, which slow down running speeds and greatly increase the probability of errors. It is therefore important that the user first consider the requirements of both memory and CPU time before embarking on a project to ascertain whether or not there is a realistic possibility of obtaining the resources to simulate [136–165].

3.2.2.2   STATISTICAL AND OTHER ERRORS

Assuming that the project can be done, there are still potential sources of error, which must be considered. These difficulties will arise in many different situations with different algorithms so we wish to mention them briefly at this time without reference to any specific simulation approach. All computers operate with limited word length and hence limited precision for numerical values of any variable. Truncation and round-off errors may in some cases lead to serious problems. In addition there are statistical errors, which mention before. What difficulties will we encounter? An inherent feature of the simulation algorithm due to the finite number of members in the 'statistical sample,' which is generated. These errors must be estimated and then a 'policy' decision must be made, that is, should more CPU time be used to reduce the statistical errors or should the CPU time available [161–165].

3.2.3   WHAT STRATEGY SHOULD FOLLOW IN APPROACHING A PROBLEM?

Most new simulations face hidden pitfalls and difficulties, which may not be apparent in early phases of the work. It is therefore often advisable to begin with a relatively simple program and use relatively small system sizes and modest running times. Sometimes there are special values of parameters for which the answers are already known (either from analytic solutions or from previous, high quality simulations) and these cases can be used to test a new simulation program. By proceeding in this manner one is able to uncover which are the parameter ranges of interest and what unexpected difficulties are present. It is then possible to refine the program and then to increase running times. Thus both CPU time and human time can be used most [165].

3.2.4 HOW DO SIMULATIONS RELATE TO THEORY AND EXPERIMENTAL WORKS?

In many cases theoretical treatments are available for models for which there is no perfect physical realization (at least at the present time). In this situation the only possible test for an approximate theoretical solution is to compare with 'data' generated from a computer simulation. As an example we wish to mention recent activity in growth models, such as diffusion-limited aggregation, for which a very large body of simulation results already exists but for which extensive experimental information is just now becoming available. It is not an exaggeration to say that interest in this field was created by simulations. Even more dramatic examples are those of reactor meltdown or large scale nuclear war: although we want to know what the results of such events would be we do not want to carry out experiments! There are also real physical systems, which are sufficiently complex that they are not presently amenable to theoretical treatment. An example is the problem of understanding the specific behavior of a system with many competing interactions and which is undergoing a phase transition. A model Hamiltonian which is believed to contain all the essential features of the physics may be proposed, and its properties may then be determined from simulations. If the simulation (which now plays the role of theory) disagrees with experiment, then a new Hamiltonian must be sought. An important advantage of the simulations is that different physical effects, which are simultaneously present in real systems, may be isolated and through separate consideration by simulation may provide a much better understanding [156–165].

3.2.5 THE ART OF RANDOM NUMBER GENERATION, BACKGROUND

Monte Carlo methods are heavily dependent on the fast, efficient production of streams of random numbers. Since physical processes, such as white noise generation from electrical circuits, generally introduce new numbers much too slowly to be effective with today's digital computers, random number sequences are produced directly on the computer using software. The use of tables of random numbers is also impractical because of the huge number of random numbers now needed for most simulations and the slow access time to secondary storage media. Since such algorithms are actually deterministic, the random number sequences, which are thus produced, are only 'pseudorandom' and do indeed have limitations which need to be understood. Thus, in the Appendix of this review, when we refer to Generation' Random numbers' programs 1 and 2, Some necessary background 'random numbers' it must be understood that we are really speaking of 'pseudorandom' numbers (Appendix A).

These deterministic features are not always negative. For example, for testing a program it is often useful to compare the results with a previous run made using exactly the same random numbers. The explosive growth in the use of Monte Carlo simulations in diverse areas of physics has prompted extensive investigation of new methods and of the reliability of both old and new techniques. Monte Carlo simulations are subject to both statistical and systematic errors from multiple sources, some of which are well understood. It has long been known that poor quality random number generation can lead to systematic errors in Monte Carlo simulation; in fact, early problems with popular generators led to the development of improved methods for producing pseudorandom numbers. As we shall show in the following discussion both the testing as well as the generation of random numbers remain important problems that have not been fully solved. In general, the random number sequences, which are needed, should be uniform, uncorrelated, and of extremely long period, that is, do not repeat over quite long intervals. In the following subsections we shall discuss several different kinds of generators. The reason for this is that it is now clear that for optimum performance and accuracy, the random number generator needs to be matched to the algorithm and computer. Indeed, the resolution of Monte Carlo studies has now advanced to the point where no generator can be considered to be completely 'safe' for use with a new simulation algorithm on a new problem. The practitioner is now faced anew with the challenge of testing the random number generator for each high-resolution application, and we shall review some of the 'tests' later in this section. The generators, which are discussed in the next subsections produce a sequence of random integers. One important topic, which we shall not consider here is the question of the implementation of random number generators on massively parallel computers. In such cases one must be certain that the random number sequences on all processors are distinct and uncorrelated. As the number of processors available to single users' increases, this question must surely be addressed, but we feel that at the present time this is a rather specialized topic and we shall not consider it further. This method for generation of random walk numbers including: Congruential method, Mixed congruential methods, Shift register algorithms Lagged Fibonacci generators Tests for quality on-uniform distributions [165, 172, 174].

3.3   MOLECULAR DYNAMICS SIMULATION

Simulations are a bridge between microscopic and macroscopic; theory and experiment. We carry out computer simulations in the hope of understanding the properties of assemblies of molecules in terms of their structure and the microscopic interactions between them. This serves as a complement to conventional experiments, enabling us to learn something new, something that cannot be found out in other ways. In this review we shall concentrate on MD. The obvious advantage of MD over MC is that it gives a route to dynamical properties of the system: transport coefficients, time-dependent responses to perturbations, rheological properties and spectra. Computer simulations act as a bridge between microscopic length and timescales and the macroscopic world of the laboratory: we provide a guess at the interactions between molecules, and obtain 'exact' predictions of bulk properties. The predictions are 'exact' in the sense that they can be made as accurate as we like, subject to the limitations imposed by our computer budget. At the same time, the hidden detail behind bulk measurements can be revealed. An example is the link between the diffusion coefficient and velocity autocorrelation function (the former easy to measure experimentally, the latter much harder). Simulations act as a bridge in another sense: between theory and experiment. We may test a theory by conducting a simulation using the same model. We may test the model by comparing with experimental results. We may also carry out simulations on the computer that are difficult or impossible in the laboratory (for example, working at extremes of temperature or pressure).Ultimately we may want to make direct comparisons with experimental measurements made on specific materials, in which case a good model of molecular interactions is essential. The aim of so-called AB initio molecular dynamics is to reduce the amount of guesswork in this process to a minimum. On the other hand, we may be interested in phenomena of a rather generic nature, or we may simply want to discriminate between good and bad theories. When it comes to aims of this kind, it is not necessary to have perfectly realistic molecular model; one that contains the essential physics may be quite suitable [122, 132, 164].

3.3.1 MILESTONES IN MD SIMULATIONS OF SWNT FORMATION

As we revisit the milestones in MD simulations of SWNT nucleation and growth, the degree to which each investigation satisfies each of these four hypothetical "tasks" will be emphasized [129].

3.3.1.1 REBO-BASED MD SIMULATIONS

The Tersoff potential and its application to hydrocarbon systems, called the REBO potential, describes bond breaking/formation using a distance-dependent many-body bond order term and correctly dissociating attractive/repulsive diatomic potentials. This approach is nevertheless based on a local molecular mechanics approach, and therefore lacks explicit descriptions of quantum phenomena and Coulomb/Vander Waals interactions. Hence, changes in the electronic structure of a dynamic, molecular system are not described. In particular, neither the evolution of conjugation and aromatic stabilization of carbons (important for sp^2-hybridized carbon), nor charge transfer effects and the near degeneracy of metal d-orbital (important for the catalyst) can be captured by MD simulations based on the Tersoff and REBO potentials. Besides these grave decencies,' the standard REBO potential has several other well-known problems. For example, in modeling gas-phase carbon densities it overestimates the sp^3 fraction compared to corresponding DFT simulations, while it underestimates the sp fraction which is of paramount importance with respect to fullerene cage self-assembly [129, 164].

3.3.1.2   CPMD SIMULATIONS

Car–Parrinello molecular dynamics (CPMD) simulations have been employed by two research groups for simulating the growth process of metal-catalyzed SWNT growth. Remarkably, the study by many researchers, was performed even before the first REBO-based MD simulations by the Maruyama group. As stated above, CPMD is a more accurate MD approach compared to REBO based MD, since it includes changes in the electronic structure throughout the dynamics. However, both reported CPMD studies were completely inadequate because of their short simulation time (due to high computational cost) and their unrealistic initial model geometries. Based on their limited simulation results, researchers suggested that the segregation of carbon linear chains and atomic rings on the surface of a liquid-like cobalt–carbide particle represents the first stage of the cap nucleation process. Researchers also simulated the incorporation of positive carbon atoms into a preassembled half-fullerene cap attached to a cobalt metal surface for 15ps. CPMD is an AB initio molecular dynamics method which is the combination of first principles electronic structure methods with MD based on Newton's equations of motion. Grand-state electronic structures were described according to DFT within plane-wave pseudo potential framework. The use of electronic structure methods to calculate the interaction potential between atoms overcomes the main shortcomings of the otherwise highly successful pair potential approach. There have been plenty of excellent reference books on MD and DFT, and some simulation tricks can be found. Here, some more details about CPMD, which are different from the traditional classical MD and DFT will be discussed. In CPMD, considering the parameters {wi}, {RI}, {av} in energy function are supposed to be time-dependent, the Lagrange:

$$E\left[\{\psi_i\},\{R_I\},\{\alpha_v\}\right]=\sum_i \int_\Omega d^3r \psi^*(r)\left[-(h^2/2m)\nabla^2\right]\psi_i(r)+U\left[n(r),\{R_I\},\{\alpha_v\}\right], \tag{1}$$

$$L=\sum_i \frac{1}{2}\mu\int_\Omega d^3r |\psi_i|^2+\sum_I \frac{1}{2}M_I R_I^2+\sum_v \frac{1}{2}\mu_v \alpha_v^2-E\left[\{\psi_i\},\{R_I\},\{\alpha_v\}\right] \tag{2}$$

where w_i are subject to the holonomic constraints:

$$\sum_i \int_\Omega d^3 r \psi_i^*(r,t) \psi_j(r,t) = \delta_{ij}. \tag{3}$$

In Eqs. (1) and (2), w_i are orbital's for electrons, RI indicate the nuclear coordinates and a_v are all the possible external constraints imposed on the system, $w^*(r)$ is the complex conjugate of wave function $w(r)$, h is the reduced Planck constant, m is the mass of electron and $n(r) = Ri \,|wi(r)\,|^2$ is the electron density; the dot indicates time derivative, M_I are the physical ionic masses. Then, the equations of motion can be written as:

$$\mu \ddot{\psi}_i(r,t) = -\frac{\delta E}{\delta \psi_i^*(r,t)} + \sum_i \Lambda_{ik} \psi_k(r,t), \tag{4}$$

$$M_I \ddot{R}_I = -\nabla_{RI} E, \tag{5}$$

$$\mu_v \ddot{\alpha}_v - -\left(\frac{\partial E}{\partial \alpha_v}\right), \tag{6}$$

That is Lagrange multipliers introduced in order to satisfy the constraints in Eq. (4). Then the equation of kinetic energy:

$$K = \sum_i \frac{1}{2} \mu \Omega d^3 r |\dot{\psi}_i|^2 + \sum_I \frac{1}{2} M_I \dot{R}_I^2 + \sum_v \frac{1}{2} \mu_v \dot{\alpha}_v^2, \tag{7}$$

is obtained. Based on the technique mentioned, CPMD extends MD beyond the usual pair-potential approximation. In addition, it also extends the application of DFT to much larger systems [129, 141, 164].

3.3.1.3 HYBRID METHODS (MD + MC)

For some complex systems Monte Carlo simulations have very low acceptance rates except for very small trial moves and hence become quite inefficient. Molecular dynamics simulations may not allow the system to develop sufficiently in time to be useful however molecular dynamics methods may actually improve a Monte Carlo investigation of the system.

A trial move is produced by allowing the molecular dynamics equations of motion to progress the system through a rather large time step. Although such a development may no longer be accurate as a molecular dynamics step, it will produce a Monte Carlo trial move which will have a much higher chance of success than a randomly chosen trial move. In the actual implementation of this method some testing is generally advisable to determine an effective value of the time step [163, 164].

3.3.1.4   AB INITIO MOLECULAR DYNAMICS

No discussion of molecular dynamics would be complete without at least a brief mention of the approach pioneered by many researchers, which combines electronic structure methods with classical molecular dynamics. In this hybrid scheme a fictitious dynamical system is simulated in which the potential energy is a functional of both electronic and ionic degrees of freedom. This energy functional is minimized with respect to the electronic degrees of freedom to obtain the potential energy surface to be used in solving for the trajectories of the nuclei. This approach has proven to be quite fruitful with the use of density functional theory for the solution of the electronic structure part of the problem and appropriately chosen pseudo potentials. These equations of motion in relation with molecular dynamics simulation can then be solved by the usual numerical methods, for example, the Verlet algorithm, and constant temperature simulations can be performed by introducing thermostats or velocity rescaling. This AB initio-method is efficient in exploring complicated energy landscapes in which both the ionic positions and electronic structure are determined simultaneously [163, 164].

3.3.1.5   DFTB-BASED MD SIMULATIONS

Neither REBO-based nor first principle-based MD simulations have succeeded in modeling the nucleation and sustained growth of a clean hexagon only SWNT from scratch as routinely achieved experimentally. This is despite the use of such tricks as injecting carbon atoms into the middle of the metal catalyst particle. It seems that one should "bite the bullet" and deal somehow with the complicated electronic structure of the evolving sp^2 network and transition metal clusters at lower computational cost. The major obstacle for such a "cheap" CPMD approach is the limited availability of metal parameters in conventional semi empirical quantum chemical methods. Most notable in this respect are the MNDO/d or PM6 codes. However, the metal parameters are typically designed for single metal atom systems and are not suitable for the treatment of metal nanoparticles. The DFTB/MD approach, which is a quantum mechanics/molecular dynamics (QM/MD) technique based on the DFTB electronic structure method, can play a role to in a gap between classical and principle MD simulations. The DFTB method is approximately two orders of magnitude faster than first principles DFT and therefore enables longer simulations and provides more adequate model systems for non-equilibrium dynamics of nano-sized clusters with quantum mechanical treatment of electrons. In addition, metal–carbon parameters for DFTB have recently been developed by the Morokuma group. An electronic temperature approach ensures the applicability of the DFTB/MD method for nanometer size metal particles with high electronic densities of states around the Fermi level, as it allows the occupancy of each molecular orbital to change smoothly from 2 to 0 depending on its orbital energy. This approach effectively incorporates the open shell nature of the system due to near-degeneracy of iron d-orbital's, as well as carbon dangling bonds. We have found that in the absence of electronic temperature the iron cluster is much less reactive. This is considered to be an artifact of the simulation of a near-closed shell electronic wave function in that case Initially, we employed a (5, 5) armchair SWNT fragment attached to an Fe_{38} cluster as a model system. The open end of the SWNT seed was terminated by hydrogen atoms, whereas the other end was bound to surface iron atoms of the Fe_{38} cluster [129, 144, 164].

3.3.1.6   SIMULATION OF ADSORBED MONO LAYERS IN SMOOTH SUBSTRATES

The study of two-dimensional systems of adsorbed atoms has attracted great attention because of the entire question of the nature of two-dimensional melting. In the absence of a periodic substrate potential, the system is free to form an ordered structure determined solely by the interparticle interactions. As the temperature is raised this planar 'solid' is expected to melt, but the nature of the transition is a matter of debate.

1.   Divide the system into sets of successively smaller Sub-simulation cells.

2. Shift the origin of the multiple expansions and calculate the multiple moments at all subcell levels starting from the lowest level.
3. Shift the origin of the local expansion and calculate the local expansion coefficients starting from the highest level.
4. Evaluate the potential and fields for each particle using local expansion coefficients for the smallest subcell containing the particle.
5. Add the contributions from other charges in the same cell and near neighbor cells by direct summation [163, 164].

3.3.1.7   PERIODIC SUBSTRATE POTENTIALS

Extensive experimental data now exist for adsorbed mono-layers on various crystalline substrates and there have been a number of different attempts made to carry out simulations which would describe the experimental observations. These falls into two general categories: lattice gas models, and off lattice models with continuous, position dependent potentials. For certain general features of the phase diagrams lattice gas models offer a simple and exceedingly efficient simulations capability. This approach can describe the general features of order or disorder transitions involving commensurate phases. An extension of the lattice gas description for the ordering of hydrogen on palladium in the C (2–2) structure has recently been proposed by giving the atoms translational degrees of freedom within a lattice cell. The situation is complicated if one wishes to consider orientation transitions involving adsorbed molecules since continuous degrees of freedom must be used to describe the angular variables. Both quadru-polar and octu-polar systems have been simulated. For a more complete description of the properties of adsorbed mono layers it is necessary to allow continuous movement of particles in a periodic potential produced by the underlying substrate. One simplification, which is often used, is to constrain the system to lie in a two-dimensional plane so that the height of the atoms above the substrate is fixed. The problem is still difficult computationally since there may be strong competition between ordering due to the atom or atom interaction and the substrate potential and in commensurate phases may result. Molecular dynamics has been used extensively for this class of problems but there have been Monte Carlo studies as well. One of the 'classic' adsorbed monolayer systems is Kr on graphite. The substrate has hexagonal symmetry with a lattice constant of 2.46A whereas the lattice constant of a compressed two-dimensional krypton solid is 1.9A. The C(1–1) structure is thus highly unfavorable and instead we find occupation of next-nearest neighbor graphite hexagons leading to a commensurate structure with lattice constant 4.26A. This means, however, that the krypton structure must expand relative to an isolated two-dimensional solid. Thus, there is competition between the length scales set by the Kr ± Kr and Kr ± graphite interactions. An important question was thus whether or not this competition could lead to an in commensurate phase at low temperatures. This is a situation in which boundary conditions again become an important consideration. If periodic boundary conditions are imposed, they will naturally tend to produce a structure, which is periodic with respect to the size of the simulation cell. In this case a more profitable strategy is to use free edges to provide the system with more freedom. The negative aspect of this choice is that finite size effects become even more pronounced. This question has been studied using a Hamiltonian and some mentioned equations [163–172].

3.3.1.8   COMPARISON OF SOME SIMULATION METHODS

As noted from the above works, the large numbers of numerical studies on CNT dynamics were based on MD simulations. In the classical MD simulations, all degrees of freedom due to the electrons are ignored, as well as quantum effects. So a more accurate description is needed, that is, potentials between electron-electron, electron-ion as well as ion-ion interactions should be considered. In this respect, the AB initio density functional theory (DFT) and CPMD were used in the study. DFT calculation, which is known as time consuming and considering more details about the electrons and the ions, can give more interactions and corrections about electron-electron, electron-ion and ion-ion, which are disregarded in classical MD. Considered these corrections, the simulated Young's moduli of CNTs would be more credible. But it can only simulate the static state. CPMD, which combines the MD simulation with quantum mechanics, treats the electronic degrees of freedom in the framework of DFT. It makes a balance between efficiency and precision, while considering the dynamic effect. However, the dynamic processes, performing in the CPMD still restricted to small system, which is a main disadvantage of CPMD (Table 3.1).

TABLE 3.1   Brief Comparison of Three Different Simulation Methods [129]

	Merits	Drawbacks
DFT	High accuracy; more details (electronic states, charge distribution, molecule orbits)	Limited to static states of small systems; slow and expensive
MD	Available for large systems; fast and cheap	Disable in chemical reaction (bond breaking/forming); empirical potentials are used which leads to low accuracy
CPMD	Combined MD with DFT; a balance between the time consuming and precision	Limited to dynamic process of small systems

It is apparent from non-hexagonal SWNT with a (n, m) chirality, cannot be achieved using REBO-based MD simulations. The effect of the increase of the conjugated electronic structure with increased stabilization is beyond the capability of simple many-body potentials. Further, the second chief objective of the metal cluster, namely keeping the carbon structure open for continued sidewall growth by incoming new carbon feedstock cannot be achieved without "cheating" (i.e., a supply of carbon from the outside of the metal cluster always results in complete encapsulation and "death of the catalyst") [164].

3.3.2 MOLECULAR INTERACTIONS

Molecular dynamics simulation consists of the numerical, step-by-step, solution of the classical equations of motion, which for a simple atomic system may be written:

$$m_i \ddot{r}_i = f_i f_i = -\frac{\partial}{\partial r_i} U \qquad (8)$$

For this purpose we need to be able to calculate the forces f i acting on the atoms, and these are usually derived from a potential energy $U(r_N)$, where $r_N = (r_1; r_2; \ldots r_N)$ represents the complete set of 3N atomic coordinates. In this section we focus on this function $U(r_N)$, restricting ourselves to an atomic description for simplicity. (In simulating soft condensed matter systems, we sometimes wish to consider non-spherical rigid units, which have rotational degrees of freedom [164, 165].

3.3.3 NON-BONDED INTERACTIONS

The part of the potential energy U non-bonded representing non-bonded interactions between atoms is traditionally split into 1-body, 2-body, 3-body terms:

$$U_{non-bonded}\left(r^N\right) = \sum_i u(r_i) + \sum_i \sum_{j>i} v\left(r_i, r_j\right) + \cdots \tag{9}$$

The u(r) term represents an externally applied potential eld or the effects of the container walls; it is usually dropped for fully periodic simulations of bulk systems. Also, it is usual to concentrate on the pair potential v $(r_i; r_j) = v (r_{ij})$ and neglect three-body (and higher order) interactions. There is an extensive literature on the way these potentials are determined experimentally, or modeled theoretically. In some simulations of complex fluids, it is sufficient to use the simplest models that faithfully represent the essential physics. In this chapter we shall concentrate on continuous, differentiable pair-potentials (although discontinuous potentials such as hard spheres and spheroids have also played a role). The Lennard-Jones potential is the most commonly used form:

$$v^{LJ}(r) = 4\varepsilon\left[\left(\frac{\sigma}{r}\right)^{12} - \left(\frac{\sigma}{r}\right)^{6}\right] \tag{10}$$

With two parameters: the diameter, and the well depth. This potential was used, for instance, in the earliest studies of the properties of liquid argon. For applications in which attractive interactions are of less concern than the excluded volume effects, which dictate molecular packing, the potential may be truncated at the position of its minimum, and shifted upwards to give what is usually termed the model. If electrostatic charges are present, we add the appropriate Coulomb potentials

$$v^{Coulomb}(r) = \frac{Q_1 Q_2}{4\pi \epsilon_0 r} \tag{11}$$

where Q_1, Q_2 are the charges and ϵ_0 is the permittivity of free space. The correct handling of long-range forces in a simulation is an essential aspect of polyelectrolyte simulations, which will be the subject of the later chapter of Holm [145, 156, 161, 164].

3.3.4   BONDING POTENTIALS

For molecular systems, we simply build the molecules out of site-site potentials. Typically, a single-molecule quantum-chemical calculation may be used to estimate the electron density throughout the molecule, which may then be modeled by a distribution of partial charges, or more accurately by a distribution of electrostatic multi poles. For molecules we must also consider inters molecular bonding interactions. The simplest molecular model will include terms of θ_{234}, and torsion angle φ_{1234}.

$$+\frac{1}{2}\sum_{\substack{bonds \\ angles}} k_{ijk}^0 \left(0_{ijk} - 0_{eq}\right)^2 \tag{12}$$

$$+\frac{1}{2}\sum_{\substack{bonds \\ angles}} k_{ijk}^0 \left(0_{ijk} - 0_{eq}\right)^2 \tag{13}$$

$$+\frac{1}{2}\sum_{\substack{torsion \\ angles}} \sum_m k_{ijkl}^{\phi,m} \left(1 + \cos(m\phi_{ijkl} - \gamma_m)\right)^2 \tag{14}$$

The geometry is illustrated in Fig. 3.1. The "bonds" will typically involve the separation $r_{ij} = \left| r_i - r_j \right|$ between adjacent pairs of atoms in a molecular framework, and we assume in Eq. (12) a harmonic form with specified equilibrium separation, although this is not the only possibility. The "bend angles" are between successive bond vectors such as and , and therefore involve three atom coordinates:

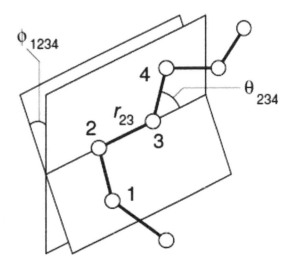

FIGURE 3.1 Geometry of a simple chain molecule, illustrating the definition of inters atomic distance r_{23}, bend angle.

$$\cos \theta_{ijk} = \hat{r}_{ij} \cdot \hat{r}_{jk} = \left(r_{ij} \cdot r_{ij} \right)^{-\frac{1}{2}} \left(r_{jk} \cdot r_{jk} \right)^{-\frac{1}{2}} \left(r_{jk} \cdot r_{jk} \right) \tag{15}$$

where $\hat{r} = r / r$. Usually this bending term is taken to be quadratic in the angular displacement from the equilibrium value, as in Eq. (15), although periodic functions are also used. The "torsion angles" ijkl are defined in terms of three connected bonds, hence four atomic coordinates:

$$\cos \phi_{ijkl} = -\hat{n}_{ijk} \cdot \hat{n}_{jkl}, where n_{ijk} = r_{ij} \times r_{jk}, n_{jkl} = r_{jk} \times r_{kl} \tag{16}$$

and $\hat{n} = n / n$, the unit normal to the plane denned by each pair of bonds. Usually the torsional potential involves an expansion in periodic functions of order m = 1,2,..., [164].

A simulation package force-ld will specify the precise form this equation. Molecular mechanics force-fields, aimed at accurately predicting structures and properties, will include many cross-terms (e.g., stretch-bend). Quantum mechanical calculations may give a guide to the "best" molecular force-eld; also comparison of simulation results with thermo physical properties and vibration frequencies is invaluable in force-eld development and renewment. A separate family of force-elds, such as AMBER, CHARMM and OPLS are geared more to larger molecules (proteins, polymers) in condensed phases; their functional form is simpler, closer to that of this equation and their parameters are typically determined by quantum chemical calculations combined with thermo physical and phase coexistence data. This eld is too broad to be reviewed here; several molecular modeling texts (albeit targeted at biological applications) should be consulted by the interested reader. The modeling of long chain molecules will be of particular interest to us, especially as an illustration of the scope for progressively simplifying and "coarse-graining" the potential model. Various explicit-atom potentials have been devised for the n-alkenes. More approximate potentials have also been constructed in which the CH_2 and CH_3 units are represented by single "united atoms." These potentials are typically less accurate and less transferable than the explicit-atom potentials, but significantly less expensive; comparisons have been made between the two approaches. For more complicated molecules this approach may need to be modified. In the liquid crystal eld, for instance, a compromise has been suggested: use the united-atom approach for hydrocarbon chains, but model phenyl ring hydrogen explicitly. In polymer simulations, there is frequently a need to economize further and coarse grain the interactions more dramatically: significantly progress has been made in recent years in approaching this problem systematically. Finally, the most fundamental properties, such as the entanglement length in a polymer melt, may be investigated using a simple chain of pseudoatoms or beads (modeled using the WCA potential and each representing several monomers), joined by an attractive finitely extensible nonlinear elastic (FENE) potential:

$$v^{FENE}(r) = \begin{cases} -\dfrac{1}{2}kR_0^2 \ln\ln\left[1-\left(\dfrac{r}{R_0}\right)^2\right] & r < R_0 (112) \\ \\ \infty & r < R_0 \end{cases}$$

(17)

The key feature of this potential is that it cannot be extended beyond, $r=R_0$, ensuring (for suitable choices of the parameters k and R_0 that polymer chains cannot move through one another [132–163].

3.3.5 PERIODIC BOUNDARY CONDITIONS

Small sample size means that, unless surface effects are of particular interest, periodic boundary conditions need to be used. Consider 1000 atoms arranged in a $10 \times 10 \times 10$ cube. Nearly half the atoms are on the outer faces, and these will have a large effect on the measured properties. Even for $10^6 = 100^3$ atoms, the surface atoms amount to 6% of the total, which is still nontrivial. Surrounding the cube with replicas of itself takes care of this problem. Provided the potential range is not too long, we can adopt the minimum image convention that each atom interacts with the nearest atom or image in the periodic array. In the course of the simulation, if an atom leaves the basic simulation box, attention can be switched to the incoming image [164].

3.3.6 NEIGHBOR LISTS

Computing the non-bonded contribution to the interatomic forces in an MD simulation involves, in principle, a large number of pair wise calculations: we consider each atom and loop over all other atoms j to calculate the minimum image separations r_{ij}. Some economies result from the use of lists of nearby pairs of atoms. Verlet suggested such a technique for improving the speed of a program. The potential cutoff sphere, of radius r_{cut}, around a particular atom is surrounded by a 'skin,' to give a larger sphere of radius r_{list}. At the rst step in a simulation, a list is constructed of all the neighbors of each atom, for which the pair separation is within r_{list}. Over the next few MD time steps, only pairs appearing in the list are checked in the force routine (Fig. 3.2) [164].

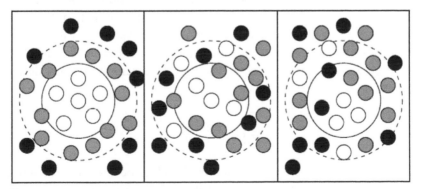

FIGURE 3.2 The Verlet lists on its construction, later, and too late. The potential cutoff range (solid circle), and the list range (dashed circle), are indicated. The list must be reconstructed before particles originally outside the list range (black) have penetrated the potential cutoff sphere.

Time to time the list is reconstructed: it is important to do this before any unlisted pairs have crossed the safety zone and come within interaction range. It is possible to trigger the list reconstruction automatically, if a record is kept of the distance traveled by each atom since the last update. The choice of list cutoff distance r_{list} is a compromise: larger listswill need to be reconstructed less frequently, but will not give as much of a saving on CPU time as smaller lists. This choice can easily be made by experimentation. For larger systems ($N \geq 1000$ or so, depending on the potential range) another technique becomes preferable. The cubic simulation box (extension to non-cubic cases is possible) is divided into a regular lattice of $n_{cell} \times n_{cell} \times n_{cell}$ cells; the first part of the method involves sorting all the atoms into their appropriate cells. This sorting is rapid, and may be performed every step. Then, within the force routine, pointers are used to scan through the contents of cells, and calculate pair forces. This approach is very efficient for large systems with short-range forces. A certain amount of unnecessary work is done because the search region is cubic, not (as for the Verlet list) spherical. Molecular dynamics evolves a nit-sized molecular configurations forward in time, in a step-by-step fashion. There are limits on the typical time scales and length scales that can be investigated and the consequences must be considered in analyzing the results. Simulation runs are typically short: typically $t \sim 10^3 - 10^6$ MD steps, corresponding to perhaps a few nanoseconds of real time, and in special cases extending to the microsecond regime. This means that we need to test whether or not a simulation has reached equilibrium before we can trust the averages calculated in it. Moreover, there is a clear need to subject the simulation averages to a statistical analysis, to make a realistic estimate of the errors. How long should we run? This depends on the system and the physical properties of interest. As the variables $a(0)$ and $a(t)$ become uncorrelated; this decay occurs over a characteristic time τ_a. Formally we may define a correlation time:

$$\tau_a = \int_0^\infty \frac{dt a(0) a(t)}{a^2} \qquad (18)$$

It is almost essential for simulation box sizes L to be large and for simulation run lengthsto be large compared with τ_a, for all properties of interest a, only then can we guarantee that reliably sampled statistical properties are obtained. Roughly speaking, the statistical error in a property calculated

as an average over a simulation run of lengthis proportional to $\sqrt{\tau_a / \tau}$:
the time average is essentially a sum of $\sim\tau/\tau_a$ independent quantities,
each an average over time τ_a. Near critical points, special care must be
taken, in that these in equalities will almost certainly not be satisfied, and
indeed one may see the onset of nonexponential decay of the correlation
functions. In these circumstances a quantitative investigation of nit size
effects and correlation times, with some consideration of the appropriate
scaling laws, must be undertaken. Phase diagrams of soft-matter systems
often include continuous phase transitions, or weakly RST-order transitions exhibiting significant pretransitional actuations. One of the most
encouraging developments of recent years has been the establishment of
reliable and systematic methods of studying critical phenomena by simulation, although typically the Monte Carlo method is more useful for this
type of study [164, 165].

3.3.7 PREDICTING GAS DIFFUSIVITY USING MOLECULAR DYNAMICS SIMULATIONS IN MOFS

3.3.7.1 EQUILIBRIUM MOLECULAR DYNAMICS SIMULATIONS IN THIS CASE STUDY

Over approximately the last decade, metal organic framework (MOF) materials have attracted a great deal of attention as a new addition to the
classes of nano porous materials. MOFs, also known as porous coordination polymers (PCPs) or porous coordination networks (PCNs) are hybrid
materials composed of single metal ions or poly nuclear metal clusters
linked by organic ligands through strong coordination bonds. Due to these
strong coordination bonds, MOFs are crystallographic ally well defined
structures that can keep their permanent porosity and crystal structure after
the removal of the guest species used during synthesis. In the literature,
MD simulations have been used to predict three different types of gas diffusivities in MOFs. These are transport diffusivity, corrected diffusivity
and self-diffusivity. MOFs have become attractive alternatives to traditional nano porous materials specifically in gas storage and gas separation
since their synthesis can be readily adapted to control pore connectivity,
structure and dimension by varying the linkers, ligands and metals in the
material. The enormous number of different possible MOFs indicates that

purely experimental means for designing optimal MOFs for targeted applications is inefficient at best. Efforts to predict the performance of MOFs using molecular modeling play an important role in selecting materials for specific applications. In many applications that are envisioned for MOFs, diffusion behavior of gases is of paramount importance. Applications such as catalysis, membranes and sensors cannot be evaluated for MOFs without information on gas diffusion rates. Most of the information on gas diffusion in MOFs has been provided by MD studies. The objective of this chapter is to review the recent advances in MD simulations of gas diffusion in MOFs. In Sections 3.3.7.2 and 3.3.7.3, the MD models used for gas molecules and MOFs will be introduced. Studies which computed single component and mixture gas diffusivities in MOFs will be reviewed in Section 3. The discussion of comparing results of MD simulations with the experimental measurements and with the predictions of theoretical correlations will be given in Sections 2.2.11, respectively. Finally, opportunities and challenges in using MD simulations for examining gas diffusion in MOFs will be summarized in Section Gas diffusion is an observable consequence of the motion of atoms and molecules as a response to external force such as temperature, pressure or concentration change. MD is a natural method to simulate the motion and dynamics of atoms and molecules. The main concept in an MD simulation is to generate successive configurations of a system by integrating Newton's law of motion. Using MD simulations, various diffusion coefficients can be measured from the trajectories showing how the positions and velocities of the particles vary with time in the system. Several different types of gas diffusion coefficients and the methods to measure them will be addressed in the next section in details. In accessing the gas diffusion in nano porous materials, equilibrium MD simulations, which model the behavior of the system in equilibrium have been very widely used. In equilibrium MD simulations, first a short grand canonical Monte Carlo (GCMC) simulation is applied to generate the initial configurations of the atoms in the nano pores. Initial velocities are generally randomly assigned to each particle (atom)based on Maxwell-Boltzmann velocity distribution. An initial NVT-MD (NVT: constant number of molecules, constant volume, constant temperature) simulation is performed to equilibrate the system. After the equilibration, Newton's equation is integrated and the positions of each particle in the system are recorded at a pre-specified rate. Nosé-Hooverther most at is very widely applied to keep the desired temperature and the integration of the system dynamics is based on the explicit N-V-T chain integrator by keeping temperature constant, Newton's equations are integrated in a canonical ensemble (NVT) instead of a micro canonical ensemble. To describe the dynamics of rigid-linear molecules such as carbon dioxide the MD algorithm is widely used. The so-called order N algorithm is implemented to calculate the diffusivities from the saved trajectories. In order to perform classical MD simulations to measure gas diffusion in MOFs' pores, force fields defining interactions between gas molecules-gas molecules and gas molecules-MOF's atoms are required. Once these force fields are specified, dynamical properties of the gases in the simulated material can be probed. These force fields will be studied in two parts: models for gas molecules (adsorbates) and models for MOFs (adsorbents) [163, 171].

3.3.7.2   MODELS FOR GASES

Diffusion of hydrogen, methane, argon, carbon dioxide and nitrogen are very widely studied in MOFs. For H_2, three different types of fluid-fluid potential models have been used. In most of the MD simulations, spherical 12–6 Lennard-Jones (LJ) model has been used for H_2. The Buch potential is known to reproduce the experimental bulk equation of state accurately for H_2. Two-site LJ models have also been used in the literature. The potential model of many researchers has been also used to account for the quadruple moment of H_2 molecules. This potential consists of a LJ core placed at the center of mass of the molecule and point charges at the position of the two protons and the center of mass [163, 171].

3.3.7.3   MODELS FOR MOFS

When the first MD simulations were performed to examine gas diffusion in MOFs at the beginning of 2004, there was no experimental data to validate the accuracy of MD studies. However, in general whenever experimental equilibrium properties such as adsorption isotherms have been reproduced by the molecular simulations, it has been observed that dynamic simulations based on the same inter atomic potentials are also reliable. Therefore, many MD studies examining gas diffusion in MOFs first showed the good agreement between experiments and simulations for gas adsorption isotherms and then used the same potential models for gas diffusion simulations. Here, it is useful to highlight that considering a wide gas loading range when comparing simulation results with experimental data is crucial. It is unreasonable to compare outcome of simulations with the experimental measurements over a very narrow range of loading and assume that good (or poor) agreement with experiment will continue to high loadings. The MD simulations have used general-purpose force fields. The MD simulations have used general-purpose force fields such as the universal force field (UFF), DREIDING force field and optimized potential for liquid simulations all-atom (OPLS-AA) force field for representing the interactions between MOF atoms and adsorbents. There are studies where the parameters of the force fields are refined to match the predictions of simulations with the experimental measurements (in most cases experimental adsorption isotherm data exist whereas experimental diffusion data do not exist) or using first principles calculation. Of course, one must be careful in refining force field parameters to match the results of simulations with the experimental data since the accuracy of the experiments are significantly affected by the defects of as synthesized MOFs or trapped residual solvent molecules present in the samples. Most MD simulations performed to date have assumed rigid MOF structures, which mean the framework atoms are fixed at their crystallographic positions. Generally, the crystallographic data for MOFs are obtained from X-ray diffraction experiments. In rigid framework simulations, only the nonbonding parameters, describing the pair wise interactions between the adsorbate and the adsorbent atoms of the particular force field, were used. It can be anticipated that the assumption of a rigid framework brings a huge computational efficiency yet the inclusion of the lattice motion and deformation is crucial for an accurate description of diffusion of large gas molecules since they fit tightly in the MOF pores, forcing the MOF to deform in order to allow migration from pore to pore. The literature summary presented so far indicates that the number of MD simulation studies with flexible MOFs and flexible force fields is very limited. More research will sure be helpful to understand the importance of lattice dynamics on diffusivity of gas molecules in MOFs. Studies to date indicated that the lattice dynamics are specifically important in computing diffusivity of large gas molecules (such as benzene) in MOFs having relatively narrow pores. Studies on flexible force fields also suggested that a force field developed for a specific MOF can be adapted to similar MOF structures (as in the case of IRMOFs) with slight modifications for doing comparative studies to provide a comprehensive understanding of gas diffusion in flexible MOFs [163, 171].

3.3.8 DUAL-CONTROL-VOLUME GRAND CANONICAL MOLECULAR DYNAMIC (DCV-GCMD) METHOD FOR PORE SIZE EFFECTS ANALYSIS ON DIFFUSION IN SWCNTS

Dual-control-volume grand canonical molecular dynamics simulations were used to study the diffusion mechanisms of counter-diffusing CH/CF_4 mixtures in cylindrical model Pores. It was found that in the Pores two different diffusion mechanisms occur independently of each other. Therefore, the pores were divided for purposes of analysis into two regions, a wall region close to the pore wall where most of the fluid molecules are located and an inner region where fewer molecules are Present but from where the main contribution to the flux comes. The dependence of the transport diffusion coefficients on pore radius and temperature were analyzed separately for these two regions. The varying contributions from fluid-fluid and fluid-wall collisions to the diffusion mechanism could be demonstrated. Whereas in the wall region surface diffusion takes Place, in the inner region diffusion occurs in the transition regime between molecular and Knudsen diffusion. Depending on different factors such as the pore size, or the temperature different diffusion mechanisms apply. There exist several simulation studies where these different diffusion mechanisms are investigated by Monte Carlo (MC) methods, by equilibrium molecular dynamics (EMD) simulations, and non-equilibrium molecular dynamics (NEMD) simulations. Several reviews exist for molecular simulation of diffusion in zealots. A simulation method that allows the direct simulation of transport diffusion is the dual-control-volume grand canonical molecular dynamic (DCV-GCMD) method. Here, the chemical potential in two control volumes is kept constant by periodically performing grand canonical Monte Carlo (GCMC) insertions and deletions. BY assigning two different values to the chemical potential in the control volumes, a gradient in the chemical potential, the driving force for transport diffusion can be established. The movement of the fluid molecules is described by MD steps where Newton's equations of motion are integrated to get a physical description of the particle movement. In general, diffusion mechanisms are classified according to the interactions of the fluid molecules with the pore wall. If the pore diameter is large in comparison to the mean free path of the fluid molecules, collisions between diffusing molecules occur far more than between the molecules and the Dore walls. The diffusion mechanism is the same as in the bulk and is called molecular diffusion. DCV-GCMD simulations of counter diffusing CH_4/CF_4 mixtures in cylindrical model pores have been carried out. The diffusion mechanisms governing transport in these pores with different radii and at different temperatures. The outline of this paper is as follows. In the following section, the simulation methods as well as the fluid model and the Dore model are introduced. In the third section, the simulation results are presented and analyzed in order to determine the underlying diffusion mechanism. DCV-GCMD simulations of binary CH_4/CF_4 mixtures were carried out in the cylindrical model. Model pore used in simulations. Each pore consists of three wall layers. The wall atoms are arranged on a quadratic grid. This allowed us to continuously vary the inner pore radius (r = 13.2 A–40A, measured from the center of the wall atoms) in a prototype system, broadly characteristic of real adsorbents, but without being constrained to, the radii of channels in known zealot structures. The model pores consisted of three coaxial cylinders built by rings forming a quadratic arid with a distance of 1.6A between individual wall atoms. The surface density of these model Pores was chosen to be 38 atoms/nm², a value close to the value of graphite (note, however, that the volume density is larger than in graphite, as the lavers were only 1.6A apart). In DCV-GCMD simulations, the Dore is divided into two control volumes and a flow region, performing a number of GCMC insertions and deletions in the control volumes, the chemical Potential is individually controlled in each control volume and a gradient in the chemical Potential can thus be established. The standard acceptance rules for GCMC insertions and deletions are used. The movement of the molecules is described by MD moves where Newton's equation of motion re integrated in order to get the trajectory. The simulation proceeds in a cyclic fashion. After a certain number of GCMC steps in the control volumes, a DCV-GCMD cycle is completed by performing MDs throughout the Dore, in the control volumes as well as in the flow region. Then the system is frozen and a new DCV-GCMD cycle starts by performing GCMC steps. It was found that in the pores two different diffusion mechanisms occurs independently surface diffusion in the layer adsorbed to the pore wall and a combination of Knudsen and molecular diffusion in the center of the pore. Therefore, the Pores were divided in two regions: a wall region and an inner region. And the diffusivities were calculated separately in each region. In the wall region formed by the first adsorbed layer, 52 to 88% of the fluid molecules are located (the Proportion depends on the Dore radius and is larger in smaller pores as here the influence of the Pore wall is larger) but only 5 to 19% of the flux takes place. The diffusivities in the wall region are about two orders of magnitude smaller than in the inner region and are of the order of magnitude of liquid-phase diffusion coefficients. The simulations at different temperatures revealed that in the wall region surface diffusion-an activated process is taking place. The simulated activations energy of 10.7 kJ/mol for CH_4 is in the order of magnitude of experimental results. For the inner region, where only 12 to 48% of the fluid molecules are located but 81 to 95% of the flux takes place, the investigations of the radius and the temperature dependence showed that diffusion is taking place in the molecular diffusion regime or in the transition regime. For pores with radius larger than approximately 23A, the diffusivity is no longer a function of the pore radius and diffusion is taking place in the molecular diffusion regime. The pore radius for which molecular diffusion starts to dominate is smaller for CF_4 than for CH_4 as the mean free path of CF_4 is smaller (CF_4 is the larger molecule). Our simulations demonstrate that is possible to use DCV-GCMD simulations to study in detail the complex diffusion processes that take place in different regions within the pore space of a porous solid [173].

3.4 SOME IMPORTANT SIMULATION ALGORITHMS THAT APPLIED IN PSD CALCULATION

3.4.1 THE MD ALGORITHM

Solving Newton's equations of motion does not immediately suggest activity at the cutting edge of research. The molecular dynamics algorithm in most common use today may even have been known to Newton. Nonetheless, the last decade has seen a rapid development in our understanding of numerical algorithms; a forthcoming review summarizes the present state of the field. Continuing to discuss, for simplicity, a system composed of atoms with coordinates $r^N = (r_1, r_2, \ldots r_N)$ and potential energy $U(r^N)$ we introduce the atomic moment a $p^N = (p_1, p_2, \ldots p_N)$ in terms of which the kinetic energy may be written $K(p^N) = \sum_{i=1}^{N} |p_i|^2 / 2m_i$, then the energy, or Hamiltonian, may be written as a sum of kinetic and potential terms $H = K + U$. Write the classical equations of motion as. This is a system of coupled ordinary differential equations. Many methods exist to perform step-by-step numerical integration of them. Characteristics of these equations are: (a) they are 'stiff," that is, there may be short and long timescales, and the algorithm must cope with both; (b) calculating the forces is expensive, typically involving a sum over pairs of atoms, and should be performed as infrequently as possible [157–160].

Also we must bear in mind that the advancement of the coordinates full two functions: (i) accurate calculation of dynamical properties, especially over times as long as typical correlation times of properties a of interest (we shall dene this later); (ii) accurately staying on the constant-energy hyper surface, for much longer times $T_{run} \gg T_a$, in order to sample the correct ensemble. To ensure rapid sampling of phase space, we wish to make the time step as large as possible consistent with these requirements. For these reasons, simulation algorithms have tended to be of low order (i.e., they do not involve storing high derivatives of positions, velocities etc.): this allows the time step to be increased as much as possible without jeopardizing energy conservation. It is unrealistic to expect the numerical method to accurately follow the true trajectory for very long times T_{run}. The 'ergodic' and 'mixing' properties of classical trajectories, that is, the fact that nearby trajectories diverge from each other exponentially quickly, make this impossible to achieve. All these observations tend to favor the Verlet algorithm in one form or another, and we look closely at this in the following section. For historical reasons only, we mention the more general class of predictor-corrector methods, which have been optimized for classical mechanical equations; further details are available elsewhere [164]:

$$\dot{r}_i = \frac{p_i}{m_i} \, and \dot{p}_i = f_i \qquad (19)$$

3.4.2   THE VERLET ALGORITHM

There are various, essentially equivalent, versions of the Verlet algorithm, including the original method and a 'leapfrog' form. Here we concentrate on the 'velocity Verlet' algorithm, which may be written:

$$r_i(t + \delta t) = r_i(t) + \delta t p_i\left(t + \frac{1}{2}\delta t\right) / m_i \tag{20}$$

$$r_i(t + \delta t) = r_i(t) + \delta t p_i\left(t + \frac{1}{2}\delta t\right) / m_i \tag{21}$$

$$p_i(t + \delta t) = p_i\left(t + \frac{1}{2}\delta t\right) + \frac{1}{2}\delta t f_i(t + \delta t) \tag{22}$$

After Eq. (21), a force evaluation is carried out, to give $f_i(t + \delta t)$ for Eq. (21). This scheme advances the coordinates and moment an over a time step δt. A piece of pseudo code illustrates how this works:

```
do step = 1, nstep
p = p + 0.5 * dt * f
r = r + dt * p/m
f = force(r)
p = p + 0.5 * dt * f
enddo
```

As we shall see shortly there is an interesting theoretical derivation of this version of the algorithm. Important features of the Verlet algorithm are: (a) it is exactly time reversible; (b) it is simplistic (to be discussed shortly); (c) it is low order in time, therefore permitting long time steps; (d) it requires just one (expensive) force evaluation per step; (e) it is easy to program [163, 164].

3.4.3   CONSTRAINTS

It is quite common practice in classical computer simulations not to attempt to represent intra molecular bonds by terms in the potential energy function, because these bonds have very high vibration frequencies (and arguably should be treated in a quantum mechanical way rather than in the classical approximation). Instead, the bonds are treated as being constrained to have fixed length. In classical mechanics, constraints are introduced through the Lagrangian or Hamiltonian formalisms. Given an algebraic relation between two atomic coordinates, for example a fixed bond length b between atoms 1 and 2, one may write a constraint equation, plus an equation for the time derivative of the constraint

$$\chi(r_1, r_2) = (r_1 - r_2)\cdot(r_1 - r_2) - b^2 = 0 \tag{23}$$

$$m_i \ddot{r}_i = f_i + \Lambda g_i \tag{24}$$

In the Lagrangian formulation, the constraint forces acting on the atoms will enter thus:

$$m_i \ddot{r}_i = f_i + \Lambda g_i \tag{25}$$

where is the undetermined multiplier and

$$g_1 = -\frac{\partial \chi}{\partial r_1} = -2(r_1 - r_2) \quad g_2 = -\frac{\partial \chi}{\partial r_2} = 2(r_1 - r_2) \tag{26}$$

It is easy to derive an exact expression for the multiplier from the above equations; if several constraints are imposed, a system of equations (one per constraint) is obtained. However, this exact solution is not what we want: in practice, since the equations of motion are only solved approximately, in discrete time steps, the constraints will be increasingly violated as the simulation proceeds. The breakthrough in this area came with the proposal to determine the constraint forces in such a way that the constraints are satisfied exactly at the end of each time step. For the original Verlet algorithm, this scheme is called SHAKE. The appropriate version of this scheme for the velocity Verlet algorithm is called RATTLE. Formally, we wish to solve the following scheme, in which we combine (r_1; r_2) into r, (p_1; p_2) into etc. for simplicity:

Choosing λ such that: $0 = \chi(r(t + \delta t))$ (27)

$$p(t + \delta t) = p\left(t + \frac{1}{2}\delta t\right) + \frac{1}{2}\delta t f(t + \delta t) + \mu g(t + \delta t) \tag{28}$$

$$p\left(t+\frac{1}{2}\delta t\right) = p(t)+\frac{1}{2}\delta t f\left(t+\delta t\right)+\lambda g(t) \tag{29}$$

$$r\left(t+\delta t\right) = r(t)+\frac{\delta t p\left(t+\frac{1}{2}\delta t\right)}{m} \tag{30}$$

Choosing such that: $\chi(t+\delta t) = \chi\left(\overline{r}(t+\delta t)+\frac{\lambda \delta t g(t)}{m}\right) = 0$ $\qquad$ (31)

The above equation may be implemented by denying unconstrained variables

$$\overline{p}\left(t+\frac{1}{2}\delta t\right) = p(t)+\frac{1}{2}\delta t f\left(t\right), \overline{r}\left(t+\delta t\right) = r(t)+\frac{\delta t \overline{p}\left(t+\frac{1}{2}\delta t\right)}{m} \tag{32}$$

Then solving the nonlinear equation for λ

$$\chi(t+\delta t) - \chi\left(\overline{r}(t+\delta t)+\frac{\lambda \delta t g(t)}{m}\right) = 0 \tag{33}$$

And substituting back

$$p\left(t+\frac{1}{2}\delta t\right) = \overline{p}\left(t+\frac{1}{2}\delta t\right)+\lambda g(t), , r\left(t+\delta t\right) = \overline{r}\left(t+\delta t\right)+\frac{\delta t \lambda g(t)}{m} \tag{34}$$

The above equation may be handled by defining

$$\overline{p}\left(t+\delta t\right) = p\left(t+\frac{1}{2}\delta t\right)+\frac{1}{2}\delta t f\left(t+\delta t\right) \tag{35}$$

Solving the equation for the second Lagrange multiplier λ,

$$\dot{\chi}(t+\delta t) = \dot{\chi}\left(r(t+\delta t), \overline{p}(t+\delta t)+\mu g(t+\delta t)\right) = 0 \tag{36}$$

(which is actually linear, since $\dot{\chi}(r,p) = -g(r)\cdot p/m$) and substituting back

$$p\left(t+\delta t\right) = \overline{p}\left(\delta t\right)+\mu g\left(t+\delta t\right) \tag{37}$$

In pseudocode this scheme may be written:

$$
\begin{aligned}
&\text{do step} = 1, \text{nstep} \\
&p = p + (dt/2) * f \\
&r = r + dt * p/m \\
&\text{lambda\_g} = \text{shake}(r) \\
&p = p + \text{lambda}_g \\
&r = r + dt * \text{lambda\_g}/m \\
&f = \text{force}(r) \\
&p = p + (dt/2) * f \\
&\text{mu\_g} = \text{rattle}(r, p) \\
&p = p + \text{mu\_g} \\
&\text{enddo}
\end{aligned}
\tag{38}
$$

The routine called shake here calculates the constraint forces λg_i necessary to ensure that the end-of-step positions r_1 satisfy Eq. (19). For a system of many constraints, this calculation is usually performed in an iterative fashion, so as to satisfy each constraint in turn until convergence; the original SHAKE algorithm was framed in this way. These constraint forces are incorporated into both the end-of-step positions and the mid-step moment a. The routine called rattle calculates a new set of constraint forces μg_i to ensure that the end-of-step momenta satisfy Eq. (20). This also may be carried out iteratively. It is important to realize that a simulation of a system with rigidly constrained bond lengths, is not equivalent to a simulation with, for example, harmonic springs representing the bonds, even in the limit of very strong springs. A subtle but crucial, difference lies in the distribution function for the other coordinates. If we obtain the conjugational distribution function by integrating over the moment a, the difference arises because in one case a set of moment a is set to zero, and not integrated, while in the other an integration is performed, which may lead to an extra term depending on particle coordinates. This is frequently called the metric tensor problem; it is explained in more detail in the references, and there are well-established ways of determining when the difference is likely to be significant and how to handle it, if necessary. Constraints also find an application in the study of rare events, or for convenience when it is desired to fix, for example, the director in a liquid crystal simulation. An alternative to constraints is to retain the intermolecular bond potentials and use a multiple time step approach to handle the fast degrees of freedom [138–163, 164].

3.4.4 ASA ALGORITHM

Although while composing each new algorithm the authors usually take
into consideration the shortcomings of previous algorithms, an ultimate
procedure for the solution of the linear Fredholm integral equation of the
first kind has not yet been elaborated. Therefore, new efforts are needed to
improve the methods for problem. First of all, the new proposed method
should be the analysis of this fast, and enable one to attain approximately
the same accuracy to extract f(H) from a measured global adsorptive isotherm we can use genetic algorithms, evolutionary algorithms, simulated
annealing, a taboo search, deterministic algorithms (our new algorithm
presented in this study, the adsorption stochastic algorithm (ASA), the
well-known relaxation (All of the powerful techniques mentioned above
minimize all unknown variables simultaneously. this algorithm can op-
timize the PSD calculating that obtained from mathematical modeling
[157–160] (Fig. 3.3).

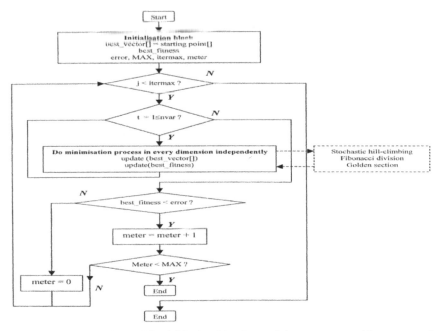

FIGURE 3.3   Flow chart of the ASA algorithm Legend: best vector—w 5(w, w, … , w),
for the starting point a uniform distribution.

3.4.5   SHN ALGORITHM

Shortcomings of conventional PSD estimation techniques. As mentioned earlier, various procedures have been used conventionally to extract the pore size distributions of various adsorbents from available adsorption isotherms. In the so-called "direct methods," a prespecified isotherm was used along with some assumed distribution (for f(r)) and then the corresponding values of (Pi) were computed by algorithm. The proposed distribution was decided to be acceptable, if and only if, the computed isotherm could re construct the experimental data. It was clearly demonstrated in our recent article that for each local adsorption isotherm (or kernel); infinite distributions (multiple solutions) can theoretically reproduce the adsorption data, while only one of them provide proper distribution for the adsorbent. Many other suboptimal solutions (which can successfully filter-out the noise and exactly recover the true under lying isotherm hidden in a set of noisy data) can be entirely inappropriate and may lead to exceedingly misleading distributions. This is not surprising, because all of these unrealistic distributions are actually the optimal solutions of a minimization problem with no definite physical meaning. Langmuir isotherm has been employed in both synthetic isotherm data generation step and PSD recovery process. This algorithm demonstrates the optimal performances of new proposed technique (SHN2) for recovery of a pre-specified true ramp-function PSD from various noisy datasets using different orders of regularizations. Figure 3.4 illustrate similar performances for single, double and triple peak Gaussians as true pore size distributions. All predictions were computed using the optimum levels of regularization (*), which were found via LOOCV method and verified manually. As it can be seen in Fig. 3.4, the new proposed method (SHN2) provides excellent prediction for pore size distributions when appropriate order of regularization with optimum regularization level have been employed. Figure 3.4 illustrates the selected performances of our newly proposed method for various PSDs using first order regularization technique at optimum levels of regularization. In all above predictions, the Halsey correlation was used for prediction of adsorbed film thickness in both isotherm generation step and PSD recovery. In all cases, Halsey correlation was used in generation steps while Harkins and Jura (Hal-Ha Ju), Deboer (Hal-dB), Micro meritics (Hal-Micr), and Kruk and Jaroniec (Hal-KJ) correlations were employed respectively in the PSD recovery operations. Evidently, the new proposed method performs adequately for all choices of adsorbed film thickness correlations. In other words, the choice of correlation used for estimation of adsorbed film thickness is not crucial [149–158].

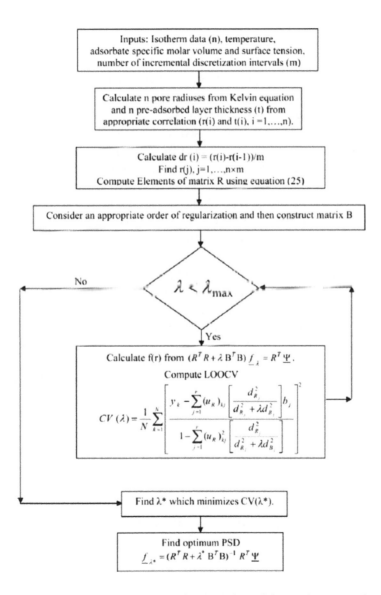

FIGURE 3.4 Flow charts for the calculation procedure of the newly proposed method (SHN2).

3.5   COMPARISON BETWEEN RECENTLY STUDY WORKS ABOUT SIMULATION AND MODELING METHODS FOR PSD CALCULATING IN CARBONS MATERIALS

Researchers description of benzene adsorption in slit-like pores with IHK and DFT methods and by ASA algorithm and GCMC simulation. In modeling the system for PSD calculating, first BET model to determine the surface area of each sample and the quantities evaluated from adsorption. The sample cell was then placed in a liquid nitrogen bath, which created an analysis temperature of approximately 77 K with nitrogen gas used as the adsorbate. The DFT was used to develop the pore size distribution data in the micro pores (<2 nm) from the nitrogen isotherm with a relative pressure range from 10^{-6} to 1, while the Barrett, Joyner, Halenda (BJH) model was used for all pores greater than 2 nm. In recent works the complexity of crystallographic and geometrical structure of many solids and their complex chemical composition are the main sources of adsorbent heterogeneity. As it was pointed out by many researchers, the main sources of surface heterogeneity are the following: different types of crystal planes, growth steps, crystal edges and corners, various atoms and functional groups exposed at the surface and available for adsorption, irregularities in crystallographic structure of surface and impurities strongly bounded with the surface. Clearly, surface heterogeneity plays generally very important role in adsorption on solids. The process of adsorption in microspores is strongly affected by the presence of small pores of different dimensions and geometrical forms. Furthermore, it is well known that uniform microspores are the source of the energetic heterogeneity in the sense that molecules adsorb in these microspores with different adsorption energies. This is caused by the change of adsorption potential field in the microspore volume. So, for micro porous solids absolute heterogeneity should be identified as a superposition of non-uniformity of micro porous structure and sources of surface heterogeneity. At the first sight the complete description of such three-dimensional complicated structure is very difficult up to the present only simplified models have been developed. Clearly, different models rely on different assumptions in order to obtain relationships allowing the calculation of the main characteristics of adsorbent structural heterogeneity. Let us consider the assumptions introduced by researchers while developed the improved HK method [126–133, 152].

1. The shape of the pores is assumed by simple slit-like. The effects of the connectivity can be neglected.
2. The molecule–surface interaction obeys the equation of Steele.
3. The adsorbed phase may be represented as one or two liquid-like layers between two parallel walls of a pore.
4. The layer thickness is the distance between two parallel planes confining centers of molecules, with the density of the layer being constant over its thickness and equal to that of the corresponding liquid.
5. The amount of the molecules of the bulk gas phase in a pore can be negligible, that is the absolute adsorption is almost equal to the surface excess adsorption [160].

Here, it is worth to point out the main weak points of the following assumptions. At first, the Steel's potential function seems to be very idealistic and far from real cases. It does not take under considerations all mentioned above sources of structural heterogeneity. Moreover, the integration to infinity of the potential function is also questionable. It is known that the number of graphite layers of the pore wall is limited to 3 or 4, corresponding to realistic pore wall thickness of about 1.1 1.5 nm. Such a simple model of adsorb ate–adsorbent interactions, also introduced in DFT, can lead to similar drawback observed on PSD calculated from a single adsorption isotherm by the DFT method. Secondly, as pointed out by different authors, the assumption of constant density inside thin layers is the most questioned. It is known that the density varies over the layer thickness. Besides the introduced simplifications, proposed mechanism of sorption in modeled uniform silt-like pores is very similar to DFT or GCMC formalism. It is important to note here, that the final form of the equation describing the grand potential functional depends on the mechanism of adsorption. Such a mechanism is related to the size of a pore. In the very small pores having one minimum of the potential in the pore center or two symmetrical minimal close to each other only one adsorb ate liquid-layer may exist. The thickness of adsorb ate layer (or adsorbed amount) depends on the relative pressure and as a consequence the local isotherm is a continuous function over a whole range of the relative pressure. In wider pores (i.e., two local minima of potential lie close the walls) the situation is more complicated. At small amount adsorbed in such type of pores two liquid layers are formed. Further increase in the relative pressure leads to the increase of the layer thickness and, consequently, to the increase of the amount adsorbed. When the distance between such two layers is small enough the mechanism of adsorption is identical to that in the small pores. So, for wider pores the first-order phase transition point should be expected [160, 161].

The criterion of first-order phase transition is the equality of the grand potential in the case of double and single layer mechanism. Detailed description of mentioned above mechanism of adsorption will be elaborated later. As a result, the mechanism of pores filling is different and depends on size of pores. In narrow pores the adsorption occurs without presetting on the pore walls (i.e., single layer mechanism). In the wider pores the formation of two thin layers on the walls is followed by the first-order phase transition in the inner volume of the pore. For this reason it is necessary to consider the pores filling mechanism separately for small pores and relatively large pores. In the case of narrow pores (small microspores) the potential energy distribution with respect to distance is characterized by one minimum located at the center of the pore or two symmetrical minima near the pore walls. As a consequence, only one liquid layer may exist in a pore. Set of equations defines the conditions of existence of a thermodynamically stable layer under the influence of an external potential field. The shape of isotherms depends on the pore width and the variation of DFs with layer thickness. This dependence should be continuous as it follows from the DFT and GCMC results. The behavior of local isotherms in small pores completely agrees with the DFT or GCMC results. The isotherms from IHK are continuous and they are shifted to the higher values of the relative pressure with increasing of effective pore widths. Effective pore width is approximated as follows, In the case of pores, which are wide enough (mesopore and some microspores), there are two local minima close to the walls. As a result, in this case, two liquid layers may exist in the same pore. Once appeared in the pore, these two liquid layers increase with the reduced pressure and the distance between them gradually decreases. As a consequence, these two liquids like layers will coalesce into one liquid-like layer if a definite value of the reduced pressure is reached and the mechanism of pores filling changes into described above single layer. For each liquid-like layer (both layers are symmetrical) the grand potential functional is defined as follows, The IHK method seems to be a promising tool for the description of adsorption in porous solids. Since DFT and GCMC methods are insufficient for the description of benzene adsorption in microspores, IHKM seems to be the most advanced and promising procedure of porosity characterization. Presented results show that the obtained pore size distribution curves depend on the type of algorithm applied for the inverting of global adsorption isotherm equation. In spite of the fact that REG leads to the smoother PSD curves than constructed by us ASA algorithm, it can sometimes generate negative parts of distribution what is without physical meaning. On the other hand, the PSD of real heterogeneous solids is rather smooth and continuous than discreet, so regularization method used in the REG algorithm seems to be natural choice (there is necessary to take under considerations the smoothing effect on the estimated PSD). Thus, this work can state that such two algorithms in connection with intuition of researchers may be successfully applied for the proper solution of the considered ill-posed problem. The results of IHK as well as ND methods show small amount of porosity in carbon blacks applied as reference materials for the construction of as plots. The comparison of the results from IHK and ND for micro porous carbons leads to the conclusion about similarity of the both methods [133, 161].

Mccallun et al. [161] represented molecular models for adsorption of water on ACs and description a comparison between simulation and experimental work results. The activated carbon is modeled as being made up of no interconnected slit pores having a distribution of pore widths given by the experimentally determined PSD. The PSD from experiment gives information on the effective or "available" pore width w. This quantity is assumed to be related to H, the distance separating the planes through the centers of the first layer of C atoms on opposing walls, N = integral 0-infinity f(w) N(w) dw, where f(w) is the PSD and N(w) is the amount adsorbed in the pores of width w at the given pressure. In principle, it is possible to calculate N by carrying out simulations to determine N(w) for a large number of fixed pore widths H covering the range of widths shown by the experimentally determined PSD. The total adsorption can then be determined from above integral, using the experimentally determined PSD, f (w). At low pressures a few pore widths suffice for such a scheme, since the isotherm changes slowly with pressure. For higher pressures a much larger number of pore widths are needed since pores filling occurs and the adsorption changes rapidly. Since the simulations are lengthy, even on the fastest supercomputers, it is not feasible at present to carry out such calculations for a large number of pore widths.

In addition in this molecular simulation method calculations were carried out using the GCMC method. This method is convenient for adsorption studies, since the chemical potential í, temperature T, and volume V are specified and kept fixed in the simulation. Since í and T are the same in the bulk and adsorbed phases at equilibrium, the thermodynamic state of the bulk phase is known in such simulations. Three types of molecular moves are attempted: molecular creation, molecular destruction, and the usual Monte Carlo translation/rotation moves. Each of these three moves is attempted with equal frequency. The type of move is chosen randomly to maintain microscopic reversibility. The probability of successful creations or destructions is strongly dependent on the density of the system. The maximum allowable rotation and displacement of a molecule are adjusted so that the combined move has an acceptance probability of about 40%. This value should ensure the most efficient probing of the phase space distribution. It is noteworthy that the values of the maximum displacement and rotation are very small compared to values encountered for nonbonding systems. In these simulations the number of adsorbed molecules fluctuates during the simulation. Calculation of the average of this number for a range of chemical potentials enables the adsorption isotherm to be constructed. The walls of the slit pores lie in the x–y plane. Normal periodic boundary conditions, together with the minimum image convention, are applied in these two directions. For low pressures, P/P_0 below 0.02, the length of the simulation cell in the two directions parallel to the walls was maintained at 10 nm for each of the pore widths studied, to maintain a sufficient number of adsorbed molecules. For higher pressures where more water molecules were present, the minimum cell length in the x and y directions was 4 nm. While this value is not high enough to require biased sampling methods, long runs are needed to ensure periodicity. Associating water molecules tend to remain in energetically favorable configurations for many MC steps, and the system thus requires many steps to reach a true equilibrium state. In our runs 500 million MC steps were used for equilibration, followed by a further 500 million steps for property averaging. Shorter runs than this were not adequate to sample desorption events. The average number of water molecules in the simulation cell varied from a few molecules at the lowest pressures to a few hundred or thousands of molecules when the pores were full; filled pores contained about 320 molecules for a pore width of 0.79 nm, 460 at 0.99 nm, 830 at 1.69 nm, and 2100 at 4.5 nm. Calculations were carried out on the Cornell Theory Center IBM SP2. In determining the adsorption isotherm, we commenced with the cell empty; a value of the fugacity corresponding to a low pressure was chosen and the average adsorption determined from the simulation. The final configuration generated at each stage was used as the starting point for simulations at higher fugacity. The pressure of the bulk gas corresponding to a given chemical potential was determined from the ideal gas equation of state. Gas phase densities corresponding to the range of chemical potentials studied were determined by carrying out simulations of the bulk gas. These were found to agree with those calculated from the ideal gas equation within the estimated errors of the simulations [161–163].

Researchers reported a review about fullerene-like models for micro porous carbons that investigated structural evolution and adsorption of fullerene-like models. Modeling the structural evolution of micro porous carbon including the formation mechanism of micro porous carbon is not well understood at the atomic level. A number of groups have attempted to model the process, and in several cases these modeling exercises have produced structures, which contain fullerene-like elements. On the starting point for the simulation was a series of all-hexagon fragments, terminated with hydrogen that the evolution of the structure showed that the H/C ratio is reduced (the temperature is increased). During this evolution, pentagons and heptagons form as well as hexagons, resulting in the formation of curved fragments. In each case the final carbon was made up of a hexagonal network with 10–15% non-hexagonal rings (pentagons and heptagons). The properties of the simulated carbons appeared to be generally consistent with experimental results. A different approach to modeling the evolutions of micro porous carbon was used. Here, the initial system consisted of carbon gas atoms at very high temperature. This choice of initial condition was intended to represent the high temperature state in a pyrolysis process after the polymer chains break down and most other elements have evaporated. The temperature was then decreased so that the atoms condensed to form a porous structure composed of curved and defected graphene sheets, in which the curvature was induced by non-hexagonal rings. In 2009, some researchers described a comprehensive molecular dynamics study of the self-assembly of carbon nano structures. The precursor for these simulations was highly disordered amorphous carbon, which was generated by rapid quenching of an equilibrated liquid sample. It was found that, under certain conditions, annealing the amorphous carbon at high temperature could lead to the highly curved sp^2 sheet structure. In modeling adsorption using fullerene-like models for micro porous carbon there have been relatively few attempts to use fullerene-like models to predict the adsorptive and other properties of micro porous carbons. By far the most ambitious programmed of work in this area has been carried out by many researchers, whose results have been published in a series of papers beginning in 2007. In the first of these 36 different carbon structures with increasing micro porosity, labeled S_0–S_{35}, were generated. Fragments were then progressively added to create the 36 structures labeled S_0–S_{35}. Pore size distribution (PSD) curves for the structures were calculated using the method of Bhattacharya and Gubbins (BG). This involves determining the statistical distribution of the radii of the largest sphere that can be fitted inside a pore at a given point. It shown that the most crowded structure, S_{35}, has a much narrower range of pore sizes than the initial S_0 structure. Argon adsorption isotherms were simulated for these structures using the parallel tempering Monte Carlo simulation method. These show that the gradual crowding of the S_0 structure (leading finally to S_{35}) leads to a decrease in the maximum number of adsorbed molecules. On the other hand, the S_0 structure exhibits less adsorption at low pressures than the more crowded ones because the average microspore diameter is larger. Also notable is the increasing sharpness of the inflection point in the isotherms, a feature, which is often reported for experimental systems. The simulated isotherms were then used to determine PSD curves, using a range of widely used methods, with the aim of checking the validity of these methods. Good agreement was found between the PSDs determined from the isotherms and the PSDs from the BG method. This confirms the validity of various methods for calculating PSD curves from adsorption data. It would also seem to confirm the validity of the fullerene-related model for micro porous carbon. The densities of these structures were calculated, and values in the range 2.18–2.24 gcm^{-3} were found, consistent with typical densities of non-graphitizing carbons. Once again, pore size distributions for the structures were determined using the BG method. Good agreement was found between the PSDs determined from the simulated adsorption data and the original PSDs from the BG method, where the PSD curve determined from the Bhattacharya-Gubbins model. The adsorption of Ne, Ar, Kr, Xe, CCl_4 and C_6H_6 on the S_0 and S_{35} carbons was modeled. The simulated data were compared with the predictions of the Dubinin–Radush kevich and Dubinin–Astakhov adsorption isotherm equations, and a good fit was found for the S_{35} carbon. For the S_0 carbon the Dubinin–Izotova (DI) equation gave a better fit because the microspores in this model have a wide distribution of diameters. The simulated isotherms exhibited a number of features similar to those seen in experimental results.

For example the isotherms for CCl_4 and C_6H_6 were temperature invariant, as observed experimentally. It was also noted that the isotherms obeyed Gurvich's rule, which states that the larger the molecular collision diameter the smaller the access to microspores, as well as other empirical and fundamental correlations developed for adsorption on micro porous carbons. The effect of oxidizing the carbon surface on porosity was analyzed. A virtual oxidation procedure was employed, in which surface carbonyls were attached to carbon atoms located on the edges of the fragments. It was assumed that the structure of the carbon skeleton remained unchanged. Pore size distributions, determined using the BG method were found not to be greatly affected by oxidation. Simulated isotherms for Ar, N_2 and CO_2 were calculated using the GCMC method. For Ar, the effect of oxidation on the isotherm was relatively small. However, for N_2 and CO_2 there were significant changes in the isotherms, due to electrostatic interactions between N_2 and CO_2 and the surface carbonyl groups. As a consequence of this, pore size distributions calculated from the simulated isotherms for N_2 and CO_2 differed markedly from those originally determined from the BG method. An important conclusion from this is that experimental PSDs determined using CO_2 (or using N_2 if there is large oxygen content) may be unreliable. A further study looked at the influence of carbon surface oxygen groups on Dubinin-Astakhov equation parameters calculated from CO_2 isotherms. It was concluded that porosity parameters calculated by fitting the DA model to experimental CO_2 adsorption data may be questionable. Others have published a number of other studies in which fullerene-like models have been used to predict the properties of micro porous carbons, but the results summarized above are sufficient to demonstrate the utility of such models method is compared with results from the Horvath–Kawazoe method [159].

KEYWORDS

- **Limited Computer Time and Memory**
- **Molecular Dynamics Simulation**
- **Monte Carlo Simulation**
- **Non-bonded Interactions**
- **The MD Algorithm**
- **The Verlet Algorithm**

CHAPTER 4

MOLECULAR MODELING AND SIMULATION

CONTENTS

4.1   INTRODUCTION

A series of advances in a variety of complementary areas cause a real progress in nanotechnology, for instance: the discoveries of atomically precise materials such as nanotubes and fullerenes, the ability of the scanning probe and the development of manipulation techniques to image and manipulate atomic and molecular configurations in real materials, the conceptualization and demonstration of individual electronic and logic devices with atomic or molecular level materials, the advances in the self-assembly of materials to be able to put together larger functional or integrated systems, and finally the advances in computational nanotechnology, physics and chemistry based modeling and simulation of possible nanomaterials, devices and applications. It turns out that at the nanoscale, devices and systems sizes have condensed sufficiently small, so that, it is possible to describe their behavior justly identically. The simulation technologies have become also predictive in nature, and many pioneer concepts and designs, which have been first proposed based on modeling and simulations, and then are followed by their realization or verification through experiments. Microscopic analysis methods are needful in order to get new functional materials and study physical phenomena on a molecular level in the nanotechnology science. These methods treat the constituent species of a system, such as molecules and fine particles. Macroscopic and microscopic quantities of interest are derived from analyzing the behavior of these species [1–3]. These approaches, called "molecular simulation methods," are represented by the Monte Carlo (MC) and molecular dynamics (MD) methods. MC methods exhibit a powerful ability to analyze thermodynamic equilibrium, but are unsuitable for investigating dynamic phenomena. MD methods are useful for thermodynamic equilibrium but are more advantageous for investigating the dynamic properties of a system in a nonequilibrium situation [4, 5].

4.2   MODELING AND SIMULATION PRINCIPLES

4.2.1   SYSTEMS

4.2.1.1   WHAT IS A SYSTEM?

A system is a set of components, which are related by some form of interaction, which act together to achieve some objective or purpose. In a system, components are the individual parts or elements that collectively make up the system and relationships are the cause-effect dependencies between components. An objective is the desired state or outcome, which the system is attempting to achieve [6].

4.2.1.2 STUDY A SYSTEM

A detailed study to determine whether, to what extent, and how automatic data-processing equipment should be used. It usually includes an analysis of the existing system and the design of the new system, including the development of system specification, which provides a basis for the selection of the equipment. There are some common steps for studying the behavior of a system [7] (Fig. 4.1).

1. Observe the behavior of a system.
2. Formulate a hypothesis about system behavior.
3. Design and carry out experiments to prove or disprove the validity of the hypothesis.
4. Often a model of the system is used
5. Measure/estimate performance.
6. Improve operation.
7. Prepare for failures.

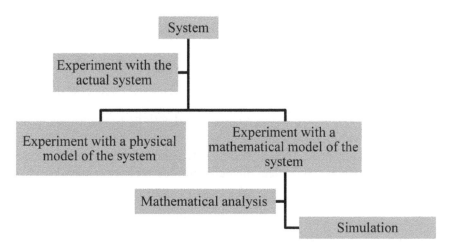

FIGURE 4.1   A schematic of a system.

4.2.1.3   AN EXAMPLE OF THE SYSTEM COMPONENTS DETERMINATION

A set of interacting components or entities operating are collected together to achieve a common goal or objective. For example in a manufacturing system its components are the machine centers, inventories, conveyor belts, production schedule and items produced and a telecommunication system is made of the messages, communication network servers [8, 9].

4.2.1.4   VARIOUS TYPES OF SYSTEMS

Systems can be classified in a variety of ways. There are natural and artificial systems, adaptive and nonadaptive systems. An adaptive system reacts to changes in its environment, whereas a nonadaptive system does not. Analysis of an adaptive system requires a description of how the environment induces a change of state. Now, some various types of system are reviewed in the following subsections.

4.2.1.4.1   NATURAL VS. ARTIFICIAL SYSTEMS

A natural system exists as a result of processes occurring in the natural world (e.g., river, universe) and an artificial system owes its origin to human activity (e.g., space shuttle, automobile)[10, 11].

4.2.1.4.2   STATIC VS. DYNAMIC SYSTEMS

A static system has structure but no associated activity (e.g., bridge, building) and a dynamic system involves time-varying behavior for complex systems (e.g., machine, U.S. economy). It deals with internal feedback loops and time delays that affect the behavior of the entire system [12, 13].

4.2.1.4.3. OPEN-LOOP VS. CLOSED-LOOP SYSTEMS

In all systems there will be an input and an output. Inputs are variables that influence the behavior of the system and outputs are variables, which determined by the system and may influence the surrounding environment. Signals flow from the input through the system and product an output.

An open-loop system cannot control or adjust its own performance but a closed-loop system controls and adjusts its own performance in response to outputs generated by the system through feedback. Feedback is the system function that obtains data on system performance (outputs), compares the actual performance to the desired performance (a standard or criterion), and determines the corrective action necessary. The controller acts on the error signal and uses the information to product the signal that actually affects the system, which we are trying to control [14] (Fig. 4.2).

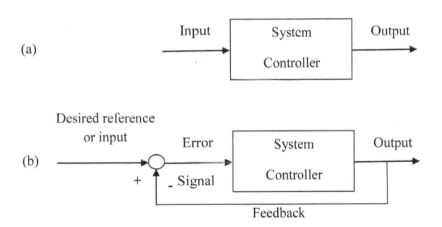

FIGURE 4.2 Graphical representations of the open-loop (a) and close-loop (b) systems.

We consider both internal and external relationships. The internal relationships connect the elements within the system, while the external relationships connect the elements with the environment, that is, with the world outside the system [15, 16].

The system is influenced by the environment through the input it receives from the environment. When a system has the capability of reacting to changes in its own state, we say that the system contains feedback. A nonfeedback, or open-loop, system lacks this characteristic.

The attributes of the system elements define its state. If the behavior of the elements cannot be predicted exactly, it is useful to take random

observations from the probability distributions and to average the performance of the objective. We say that a system is in equilibrium or in the
steady state if the probability of being in some state does not vary in time.
There are still actions in the system, that is, the system can still move from
one state to another, but the probabilities of its moving from one state to
another are fixed. These fixed probabilities are limiting probabilities that
are realized after a long period of time, and they are independent of the
state in which the system started. A system is called stable if it returns to
the steady state after an external shock in the system. If the system is not
in the steady state, it is in a transient state [15].

4.3   MODELS

The first step in studying a system is building a model. A model is presentation of an object, a system, or an idea in some form other than that of the
entity itself. A critical step in building the model is constructing the objective function, which is a mathematical function of the decision variables.
For the complex systems a model describes the behavior of systems by
using the construct theories or hypotheses, which could be accounted for
the observed behavior. So the model can predict future behavior and the
effects that will be produced by changes in the system due to the analysis
of the proposed systems.

4.3.1   MODELING APPROACH

Computational models, which also called simulation models used to design new systems study and improve the behavior of existing systems.
They allow the use of an interactive design methodology (sometimes
called computational steering) so used in most branches of science and
engineering (Fig. 4.3).

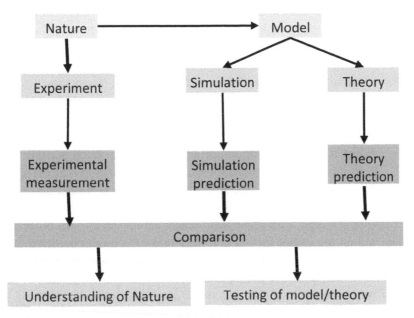

FIGURE 4.3   Modeling and experimental.

There are many types of models such as:

1. A scale model of the real system, for example, a model aircraft in a wind tunnel or a model railway [17].
2. A physical model in different physical system to the real one, for example, colored water in tubes has been used to simulate the flow of coal in a mine. More common in the use of electrical circuits–analog computers are based on this idea [18].
3. Mathematical Model: A description of a system where the relationship between variables of the system are expressed in a mathematical form [19, 20].
4. Deterministic vs. stochastic models: In deterministic models, the input and output variables are not subject to random fluctuations, so that the system is at any time entirely defined by the initial conditions and in stochastic models, at least one of the input or output variables is probabilistic or involves randomness [21, 22].

Among the models mathematical models are the most applicable models, which have many advantages such as [23]:

1. Enable the investigators to organize their theoretical beliefs and empirical observations about a system and to deduce the logical implications of this organization.
2. Lead to improved system understanding.
3. Bring into perspective the need for detail and relevance.
4. Expedite the analysis.
5. Provide a framework for testing the desirability of system modifications.
6. Allow for easier manipulation than the system itself permits.
7. Permit control over more sources of variation than a direct study.
8. An additional advantage is that a mathematical model describes a problem more concisely than a verbal description does.

To build a model, important factors that act on the system must be included and un important factors that only make the model harder to build, understand, and solve should be omitted. For a continuous model, a set of equations that describe the behavior of a system as a continuous function of time t are written. Models use statistical approximations for systems that cannot be modeled using precise mathematical equations. While building a model it must be taken to ensure that it remains a valid representation of the problem. In order to get this purpose, a scientific model necessarily embodies elements of two conflicting attributes-realism and simplicity [24].

4.3.2   COMPUTATIONAL MODELS, ACCURACY, AND ERRORS

Appropriate balance between accuracy and complexity must be obtained in a model. An accurate representation of the physical system must be simple enough to implement as a program and solve on a computer in a reasonable amount of time. On the one hand, the model should serve as a reasonably close approximation to the real system and incorporate most of the important aspects of the system. On the other hand, the model must not be so complex that it is impossible to understand and manipulate. Adding details to the model makes the solution more difficult and converts the method for solving a problem from an analytical to an approximate numerical one [25].

In addition, it is not obligatory for the model to approximate the system to demonstrate the measure of effectiveness for all various alternatives. It needs to be a high correlation between the prediction by the model and what would actually happen with the real system. To specify whether this requirement is satisfied or not, it is important to test and establish control over the solution.

Usually, the model by reexamining the formulation of the problem and revealing possible flaws must be tested. Another touchstone for judging the validity of the model is determining whether all mathematical expressions are dimensionally consistent. A third useful test consists of varying input parameters and checking that the output from the model behaves in a plausible manner. The fourth test is the so-called retrospective test. It involves using historical data to reconstruct the past and then determining how well the resulting solution would have performed if it had been used. Comparing the effectiveness of this hypothetical performance with what actually happened then indicates how well the model predicts the reality. However, a disadvantage of retrospective testing is that it uses the same data that guided formulation of the model. Unless the past is a true replica of the future, it is better not to resort to this test at all.

Suppose that the conditions under which the model was built change. In this case the model must be modified and control over the solution must be established. Often, it is desirable to identify the critical input parameters of the model, that is, those parameters subject to changes that would affect the solution, and to establish systematic procedures to control them. This can be done by sensitivity analysis, in which the respective parameters are varied over their ranges to determine the degree of variation in the solution of the model [25].

After constructing a mathematical model for the problem under consideration, the next step is to derive a solution from this model. There are analytic and numerical solution methods. An analytic solution is usually obtained directly from its mathematical representation in the form of formula.

A numerical solution is generally an approximate solution obtained as a result of substitution of numerical values for the variables and parameters of the model. Many numerical methods are iterative, that is, each successive step in the solution uses the results from the previous step [26].

4.4   SIMULATION

The process of conducting experiments on a model of a system in lieu of either (i) direct experimentation with the system itself, or (ii) direct analytical solution of some problem associated with the system. A simulation of a system is the operation of a model of the system, as an imitation of the real system. A tool to evaluate the performance of a system, existing or proposed, under different configurations of interest and over a long period of time so a simulation of an industrial process is to learn about its behavior under different operating conditions in order to improve the process. Simulation is indeed an invaluable and very versatile tool in those problems where analytic techniques are inadequate. Simulation is a numerical technique for conducting experiments on a digital computer, which involves certain types of mathematical and logical models that describes the behavior of a business or economic system (or some component thereof) over extended periods of real time. Simulation does not require that a model be presented in a particular format. It permits a considerable degree of freedom so that a model can bear a close correspondence to the system being studied. The results obtained from simulation are much the same as observations or measurements that might have been made on the system itself [7, 27].

4.4.1   REASONS FOR SIMULATION

Simulation allows experimentation, although computer simulation requires long programs of some complexity and is time consuming. Yet what are the other options? The answer is direct experimentation or a mathematical model. Direct experimentation is costly and time consuming, yet computer simulation can be replicated taking into account the safety and legality issues. On the other hand, one can use mathematical models yet mathematical models cannot cope with dynamic effects. Also, computer simulation can sample from nonstandard probability distribution. One can summarize the advantages of computer simulation [28, 29]:

Simulation, first, allows the user to experiment with different scenarios and, therefore, helps the modeler to build and make the correct choices without worrying about the cost of experimentations. The second reason for using simulation is the time control. The modeler or researcher can expand and compress time, just like pressing a fast forward button on life. The third reason is like the rewind button: seeing a scene over and over will definitely shed light on the answer of the question, "why did this happen?"

The fourth reason is "exploring the possibilities." Considering that the package user would be able to witness the consequences of his/her actions on a computer monitor and, as such, avoid jeopardizing the cost of using the real system; therefore, the user will be able to take risks when trying new things and diving in the decision pool with no fears hanging over her/his neck.

As in chess, the winner is the one who can visualize more moves and scenarios before the opponent does. In business the same idea holds. Making decisions on impulse can be very dangerous, yet if the idea is envisaged on a computer monitor then no harm is really done, and the problem is diagnosed before it even happens. Diagnosing problems is the fifth reason why people need to simulate [30].

Likewise, the sixth reason tackles the same aspect of identifying constraints and predicting obstacles that may arise, and is considered as one major factor why businesses buy simulation software. The seventh reason addresses the fact that many times decisions are made based on "someone's thought" rather than what is really happening.

When studying some simulation packages, the model can be viewed in 3-D. This animation allows the user "to detect design flaws within systems that appear credible when seen on paper or in a 2-D CAD drawing." The ninth incentive for simulation is to "visualize the plan."

It is much easier and more cost effective to make a decision based on predictable and distinguished facts. Yet, it is a known fact that such luxury is scarce in the business world [28].

Nevertheless, before trying out the "what if" scenario many would rather have the safety net beneath them. Therefore, simulation is used for "preparing for change" [29].

In addition, the 13th reason is evidently trying different scenarios on a simulated environment; proving to be less expensive, as well as less disturbing, than trying the idea in real life. Therefore, simulation software does save money and effort, which denotes a wise investment. In any field listing the requirements can be of tremendous effort, for the simple reason that there are so many of them. As such, the 14th reason crystallizes in avoiding overlooked requirements and imagining the whole scene, or the trouble of having to carry a notepad to write on it when remembering a forgotten requirement. While these recited advantages are of great significance, yet many disadvantages still show their effect, which are also summarized in Refs. [28, 29, 31].

4.4.2   DANGERS OF SIMULATION

Becoming too enthusiastic about a model and forget about the experimental frame. Force reality into the constraints of a model and forget the model's level of accuracy. Also it should not be forgotten that all models have simplifying assumptions. If two modelers work together and cannot agree on a model, which can be due to the human nature, a consequence of it is the difficulty of interpreting the results of the simulation. It is the simple fact that simulation is not the solution for all problems. Hence, certain types of problems can be solved using mathematical models and equations.

Simulation may be used in appropriately. Simulation is used in some cases when an analytical solution is possible, or even preferable. This is particularly true in the case of small queuing systems and some probabilistic inventory systems, for which closed form models (equations) are available.

Although of all dangerous of simulation, recent advances in simulation methodologies, availability of software, and technical developments have made simulation one of the most widely used and accepted tools in system analysis and operations research [32].

4.4.3   MODEL TRAINING

Simulation models can provide excellent training when designed for that purpose. Used in this manner, the team provides decision inputs to the simulation model as it progresses. The team, and individual members of the team, can learn from their mistakes, and learn to operate better. Moreover, training any team using a simulated environment is less expensive than real life. Some of model building requires special training. Model building is an art that is learned over time and through experience. Furthermore, if two models of the same system are constructed by two competent individuals, they may have similarities, but it is highly unlikely that they will be identical [33].

4.4.4   SIMULATION APPROACHES

There are four significant simulation approaches or methods used by the simulation community:
1. Process interaction approach.
2. Event scheduling approach.
3. Activity scanning approach.
4. Three-phase approach.

There are other simulation methods, such as transactional-flow approach, that are known among the simulation packages and used by simulation packages like Pro Model, Arena, Extend, and Witness. Another, known method used specially with continuous models is stock and flow method [34–39].

4.4.5   PROCESS INTERACTION APPROACH

The simulation structure that has the greatest intuitive appeal is the process interaction method. In this method, the computer program emulates the flow of an object (for example, a load) through the system. The load moves as far as possible in the system until it is delayed, enters an activity, or exits from the system. When the load's movement is halted, the clock advances to the time of the next movement of any load.

This flow, or movement, describes in sequence all of the states that the object can attain in the system. In a model of a self-service laundry, for example, a customer may enter the system, wait for a washing machine to become available, wash his or her clothes in the washing machine, wait for a basket to become available, unload the washing machine, transport the clothes in the basket to a dryer, wait for a dryer to become available, unload the clothes into a dryer, dry the clothes, unload the dryer, and then leave the laundry. Each state and event is simulated. Process interaction approach is used by many commercial packages [40].

4.4.6   TRANSACTION FLOW APPROACH

Transaction flow approach was first introduced by GPSS in 1962. Transaction flow is a simpler version of process interaction approach, as the following clearly states: world-view was a cleverly disguised form of process interaction that put the process interaction approach within the grasp of ordinary users." In transaction flow approach models consist of entities (units of traffic), resources (elements that service entities), and control elements (elements that determine the states of the entities and resources). Discrete simulators, which are generally, designed for simulating detailed processes, such as call centers, factory operations, and shipping facilities, rely on such approach [40–42].

4.4.7   EVENT SCHEDULING APPROACH

The basic concept of the event scheduling method is to advance time to the moment when something happens next (i.e., when one event ends, time is advanced to the time of the next scheduled event). An event usually releases a resource. The event then reallocates available objects or entities by scheduling activities, in which they can now participate. For example, in the self-service laundry, if a customer's washing is finished and there is a basket available, the basket could be allocated immediately to the customer, who would then begin unloading the washing machine. Time is advanced to the next scheduled event (usually the end of an activity) and activities are examined to see whether any can now start as a consequence. Event scheduling approach has one advantage and one disadvantage as: "the advantage was that it required no specialized language or operating system support. Event-based simulations could be implemented in procedural languages of even modest capabilities."While the disadvantage "of the event-based approach was that describing a system as a collection of events obscured any sense of process flow." As such, "in complex systems, the number of events grew to a point that following the behavior of an element flowing through the system became very difficult" [35, 39, 40].

4.4.8 ACTIVITY SCANNING APPROACH

Another simulation modeling structure is activity scanning. Activity scanning is also known as the two-phase approach. Activity scanning produces a simulation program composed of independent modules waiting to be executed. In the first phase, a fixed amount of time is advanced, or scanned. In phase two, the system is updated (if an event occurs). Activity scanning is similar to rule-based programming (if the specified condition is met, then a rule is executed) [36, 40].

4.4.9 SIMULATION ASPECTS

Three aspects need to be considered when planning a computer simulation project:
1. Time-flow handling
2. Behavior of the system
3. Change handling

The flow of time in a simulation can be handled in two manners: the first is to move forward in equal time intervals. Such an approach is called time-slice. The second approach is next-event which increments time in variable amounts or moves the time from state to state. On one hand, there is less information to keep in the time-slice approach. On the other hand, the next-event approach avoids the extra checking and is more general [40, 43].

The behavior of the system can be deterministic or stochastic: deterministic system, of which its behavior would be entirely predictable, whereas, stochastic system, of which its behavior cannot be predicted but some statement can be made about how likely certain events are to occur.

The change in the system can be discrete or continuous. Variables in the model can be thought of as changing values in four ways [43]:
1. Continuously at any point of time: thus, change smoothly and values of variables at any point of time.
2. Continuously changing but only at discrete time events: values change smoothly but values accessible at predetermined time.
3. Discretely changing at any point of time: state changes are easily identified but occur at any time.

4. Discretely changing at any point of time: state changes can only occur at specified point of time.

Others define 3 and 4 as discrete event simulation as follows: "a discrete-event simulation is one in which the state of a model changes at only a discrete, but possibly random, set of simulated time points." Mixed or hybrid systems with both discrete and continuous change do exist. Actually simulation packages try to include both.

In the natural sciences one makes to model the complex processes occurring in nature as accurately as possible. The first step in this direction is the description of nature. It serves to develop an appropriate system of concepts. However, in most cases, only observation is not enough to find the underlying principles. Most processes are too complex and cannot be clearly separated from other processes that interact with them. Instead, if it is possible, the scientist creates the conditions under the process is to be observed in an experiment. This method allows discovering how the observed event depends on the chosen conditions and allows inferences about the principles underlying the behavior of the observed system. The goal is the mathematical formulation of the underlying principles by using a theory of the phenomena under investigation and describes how certain variables behave independence of each other and how they change under certain conditions over time. This is mostly done by means of differential and integral equations. The resulting equations, which encode the description of the system or process, are referred to as a mathematical model.

A model that has been confirmed does not only permit the precise description of the observed processes, but also allows the prediction of the results of similar physical processes within certain bounds. Thereby, experimentation, the discovery of underlying principles from the results of measurements, and the translation of those principles into mathematical variables and equations go hand in hand. Theoretical and experimental approaches are therefore most intimately connected. The phenomena that can be investigated in this way in physics and chemistry extend over very different orders of magnitudes. They can be found from the smallest to the largest observable length scales, from the investigation of matter in quantum mechanics to the study of the shape of the universe. The occurring dimensions range from the nanometer range (10^{-9} m) in the study of properties of matter on the molecular level to 10^{23} meters in the study of galaxy clusters. Similarly, the time scales that occur in these models (i.e., the typical time intervals in which the observed phenomena take place) are vastly different. They range in the mentioned examples from 10^{-12} or even 10^{-15} sec to 10^{17} sec, thus from picoseconds or even femtoseconds up to time intervals of several billions of years. The masses occurring in the models are just as different, ranging between 10^{-27} kilograms for single atoms to 10^{40} kilograms for entire galaxies.

The wide range of the described phenomena shows that experiments cannot always be conducted in the desired manner. For example in astrophysics, there are only few possibilities to verify models by observations and experiments and to thereby confirmed them, or in the opposite case to reject models, to falsify them. On the other hand, models that describe nature sufficiently well are often so complicated that no analytical solution can be found.

Take for example the case of the Vander Waals equation to describe dense gases or the Boltzmann equation to describe the transport of rarefied gases. Therefore, one usually develops a new and simplified model that is easier to solve. However, the validity of this simplified model is in general more restricted. To derive such models one often uses techniques such as averaging methods, successive approximation methods, matching methods, asymptotic analysis and homogenization. Unfortunately, many important phenomena can only be described with more complicated models. But then these theoretical models can often only be tested and verified in a few simple cases. As an example consider again planetary motion and the gravitational force acting between the planets according to Newton's law. As is known, the orbits following from Newton's law can be derived in closed form only for the two-body case. For three bodies, analytical solutions in closed form in general no longer exist. This is also true for our planetary system as well as the stars in our galaxy.

Many models, for example in materials science or in astrophysics, consist of a large number of interacting bodies (called particles), as for example atoms and molecules. In many cases the number of particles can reach several millions or more. But large numbers of particles do not only occur on a microscopic scale. These are some of the reasons why computer simulation has recently emerged as a third way in science besides the experimental and theoretical approach. Over the past years, computer simulation has become an indispensable tool for the investigation and prediction of physical and chemical processes. In this context, computer simulation means the mathematical prediction of technical or physical processes on modern computer systems [44].

The following procedure is typical in this regard: A mathematical-physical model is developed from observation. The derived equations, in most cases, valid for continuous time and space, are considered at selected discrete points in time and space. For instance, when discretizing in time, the solution of equations is no longer to be computed at all points in time, but is only considered at selected points along the time axis. Differential operators, such as derivatives with respect to time, can then be approximated by difference operators. The solution of the continuous equations is computed approximately at those selected points. The more densely those points are selected, the more accurately the solution can be approximated. Here, the rapid development of computer technology, which has led to an enormous increase in the computing speed and the memory size of computing systems, now allows simulations that are more and more realistic. The results can be interpreted with the help of appropriate visualization techniques. If corresponding results of physical experiments are available, then the results of the computer simulation can be directly compared. This leads to a verification of the results of the computer simulation or to an improvement in the applied methods or the model (for instance by appropriate changes of parameters of the model or by changing the used equations).

Altogether, for a computer experiment, one needs a mathematical model. But the solutions are now obtained approximately by computations, which are carried out by a program on a computer. This allows studying models that are significantly more complex and therefore more realistic than those accessible by analytical means. Furthermore, this allows avoiding costly experimental setups. In addition, situations can be considered that otherwise could not be realized because of technical shortcomings or because they are made impossible by their consequences. For instance, this is the case if it is hard or impossible to create the necessary conditions in the laboratory, if measurements can only be conducted under great difficulties or not at all, if experiments would take too long or would run too fast to be observable, or if the results would be difficult to interpret. In this way, computer simulation makes it possible to study phenomena not accessible before by experiment. If a reliable mathematical model is available that describes the situation at hand accurately enough, it does in general not make a difference for the computer experiment. Obviously this is different, if the experiment would actually have to be carried out in reality. Moreover, the parameters of the experiment can easily be changed.

And the behavior of solutions of the mathematical model with respect to such parameters changes can be studied with relatively little effort [44] (Fig. 4.4).

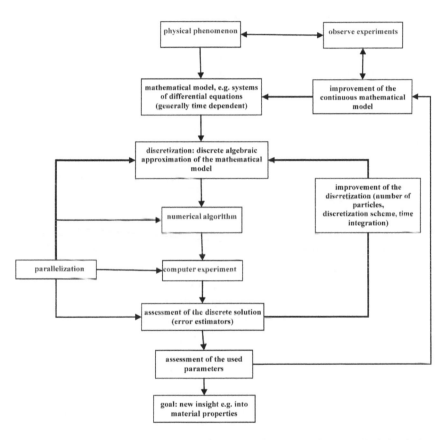

FIGURE 4.4 Schematic presentation of the typical approach for numerical simulations.

In nanotechnology numerical simulation can help to predict properties of new materials that do not yet exist in reality. And it can help to identify the most promising or suitable materials. The trend is towards virtual laboratories in which materials are designed and studied on a computer. Simulation offers the possibility of determining mean or average properties for the material macroscopic characterization. At whole it can be said, computer experiments act as a link between laboratory experiments and mathematical- physical theory.

Each of the partial steps of a computer experiment must satisfy a number of requirements. First and foremost, the mathematical model should describe reality as accurately as possible. In general, certain compromises between accuracy in the numerical solution and complexity of the mathematical model have to be accepted. In most cases, the complexity of the models leads to enormous memory and computing time requirements, especially if time-dependent phenomena are studied. Depending on the formulation of the discrete problem, several nested loops have to be executed for the time dependency, for the application of operators, or also for the treatment of nonlinearities.

Current researches therefore have focus in particular on the development of methods and algorithms that allow to compute the solutions of the discrete problem as fast as possible (multilevel and multi scale methods, multiple methods, fast Fourier transforms) and that can approximate the solution of the continuous problem with as little memory as possible. More realistic and therefore in general more complex models require faster and more powerful algorithms. Another possibility to run larger problems is the use of vector computers and parallel computers. Vector computers increase their performance by processing similar arithmetical instructions on data stored in a vector in an assembly line-like fashion. In parallel computers, several dozen in many thousands of powerful processors. The processors in use today have mostly a RISC (reduced instruction set computer) processor architecture. They have fewer machine instructions compared to older processors, allowing a faster, assembly line-like execution of the instructions which are assembled into one computing system [45]. These processors can compute concurrently and independently and can communicate with each other to improve portability of programs among parallel computers from different manufacturers and to simplify the assembly of computers of various types to a parallel computer, which has uniform standards for data exchange between computers. A reduction of the required computing time for a simulation is achieved by distributing the necessary computations to several processors. Up to a certain degree, the computations can then be executed concurrently. In addition, parallel computer systems in general have a substantially larger main memory than sequential computers. Hence, larger problems can be treated [44].

Figure 4.5 shows the development of the processing speed of high performance. The performances in flop/s is plotted versus the year, for the fastest parallel computer in the world, and for the computers at position 100 and 500 in the list of the fastest parallel computers in the world. Personal computers and workstations have seen a similar development of their processing speed. Because of that, satisfactory simulations have become possible on these smaller computers [44].

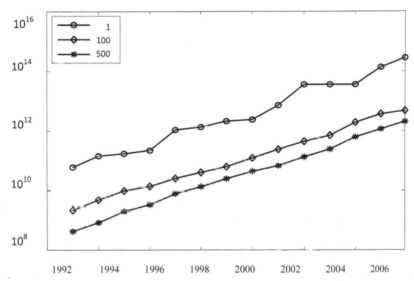

FIGURE 4.5 Development of processing speed over's the last years (parallel Linpack benchmark); fastest (1) 100th fastest (100) and 500th fastest (500) computer in the world; up to now the processing speeds increases tenfold about every four years.

4.5   PARTICLE MODELS

An important area of numerical simulation deals with so-called particle models. These are simulation models in which the representation of the physical system consists of discrete particles and their interactions. For instance, systems of classical mechanics can be described by the positions, velocities, and the forces acting between the particles. In this case the particles do not have to be very small either with respect to their dimensions or with respect to their mass, as possibly suggested by their name. Rather they are regarded as fundamental building blocks of an abstract model. For this reason the particles can represent atoms or molecules. The particles carry properties of physical objects, as for example mass, position, velocity or charge. The state and the evolution of the physical system is represented by these properties of the particles and by their interactions, respectively [46]. The use of Newton's second law results in a system of ordinary differential equations of second order describing how the acceleration of any particle depends on the force acting on it. The force results from the interaction with the other particles and depends on their position. If the positions of the particles change relative to each other, then in general also the forces between the particles change. The solution of the system of ordinary differential equations for given initial values then leads to the trajectories of the particles. This is a deterministic procedure, meaning that the trajectories of the particles are in principle uniquely determined for all times by the given initial values.

But why is it reasonable at all to use the laws of classical mechanics when at least for atomic models the laws of quantum mechanics should be used? Should not Schrodinger's equation be employed as equation of motion instead of Newton's laws? And what does the expression "interaction between particles" actually mean, exactly?

If one considers a system of interacting atoms, which consists of nuclei and electrons, one can in principle determine its behavior by solving the Schrodinger equation with the appropriate Hamilton operator. However, an analytic or even numerical solution of the Schrodinger equation is only possible in a few simple special cases. Therefore, approximations have to be made. The most prominent approach is the Born-Oppenheimer approximation. It allows a separation of the equations of motions of the nuclei and of the electrons. The intuition behind this approximation is that the significantly smaller mass of the electrons permits them to adapt to the new position of the nuclei almost instantaneously. The Schrodinger equation for the nuclei is therefore replaced by Newton's law. The nuclei are then moved according to classical mechanics, but using potentials that result from the solution of the Schrodinger equation for the electrons. For the solution of this electronic Schrodinger equation approximations have to be employed. Such approximations are for instance derived with the Hartree-Fock approach or with density functional theory. This approach is known as AB initio molecular dynamics. However, the complexity of the model and the resulting algorithms enforces a restriction of the system size to a few thousand atoms [46].

A further simplification is the use of parameterized analytical potentials that just depend on the position of the nuclei (classical molecular dynamics). The potential function itself is then determined by fitting it to the results of quantum mechanical electronic structure computations for a few representative model configurations and subsequent force-matching or by fitting to experimentally measured data. The use of these very crude approximations to the electronic potential hyper-surface allows the treatment of systems with many millions of atoms. However, in this approach quantum mechanical effects are lost to a large extent.

4.5.1 PHYSICAL SYSTEMS FOR PARTICLE MODELS

The following list gives some examples of physical systems that can be represented by particle systems in a meaningful way. They are therefore amenable to simulation by particle methods [47]:

4.5.1.1 SOLID STATE PHYSICS

The simulation of materials on an atomic scale is primarily used in the analysis of known materials and in the development of new materials. Examples for phenomena studied in solid state physics are the structure conversion in metals induced by temperature or shock, the formation of cracks initiated by pressure, shear stresses, etc. in fracture experiments, the propagation of sound waves in materials, the impact of defects in the structure of materials on their load-bearing capacity and the analysis of plastic and elastic deformations [48].

4.5.1.2 FLUID DYNAMICS

Particle simulation can serve as a new approach in the study of hydrodynamical instabilities on the microscopic scale, as for instance, the Rayleigh-Taylor or Rayleigh-Benard instability. Furthermore, molecular dynamics simulations allow the investigation of complex fluids and fluids mixtures, as for example emulsions of oil and water, but also of crystallization and of phase transitions on the microscopic level [49].

4.5.1.3 BIOCHEMISTRY

The dynamics of macromolecules on the atomic level is one of the most prominent applications of particle methods. With such methods it is possible to simulate molecular fluids, crystals, amorphous polymers, liquid crystals, zeolites, nuclear acids, proteins, membranes and many more biochemical materials [50].

4.5.1.4   ASTROPHYSICS

In this area, simulations mostly serve to test the soundness of theoretical models. In a simulation of the formation of the large-scale structure of the universe, particles correspond to entire galaxies. In a simulation of galaxies, particles represent several hundred to thousand stars. The force acting between these particles results from the gravitational potential [46, 51].

4.5.2   COMPUTER SIMULATION OF PARTICLE MODELS

In the computer simulation of particle models, the time evolution of a system of interacting particles is determined by the integration of the equations of motion. Here, one can follow individual particles, see how they collide, repel each other, attract each other, how several particles are bound to each other, are binding to each other, or are separating from each other. Distances, angles and similar geometric quantities between several particles can also be computed and observed over time. Such measurements allow the computation of relevant macroscopic variables such as kinetic or potential energy, pressure, diffusion constants, transport coefficient, structure factors, spectral density functions, distribution functions, and many more. In most cases, variables of interest are not computed exactly in computer simulations, but only up to certain accuracy. Because of that, it is desirable to achieve:

- · an accuracy as high as possible with a given number of operations,
- · a given accuracy with as few operations as possible, or
- · a ratio of effort (number of operations) to achieved accuracy, which is as small as possible.

Clearly the last alternative includes the first two as special cases. A good algorithm possesses a ratio of effort (costs, number of operations, necessary memory) to benefit (achieved accuracy) that is as favorable as possible. As a measure for the ranking of algorithms one can use the quotient:

This is a number that allows the comparison of different algorithms. If it is known how many operations are minimally needed to achieve certain accuracy, this number shows how far a given algorithm is from optimal. The minimal number of operations to achieve a given accuracy ε is called ε-complexity. The -complexity is thus a lower bound for the number of operations for any algorithm to achieve an accuracy of ε [52].

4.7   HIGH LEVEL ARCHITECTURE (HLA)

In 2000, HLA became an IEEE standard for distributed simulation. It consists of several federates (members of the simulation), that make up a federation (distributed simulation), work together and use a common runtime infrastructure (RTI). The RTI interface specification, together with the HLA, [53] object model template (OMT) and the HLA rules, are the key defining elements of the whole architecture [53–55] (Table 4.1).

TABLE 4.1   Continuous and Discrete Simulations

Continuous	Discrete
Continuously advances time and system state.	System state changes only when events occur.
Time advances in increments small enough to ensure accuracy.	Time advances from event to event.
State variables updated at each time step.	State variables updated as each event occurs.

4.8   HLA OBJECT MODELS

Whereas the interface specification is the core of the transmission system that connects different software systems, regardless of platform and language, the object model template is the language spoken over that line.

HLA has an object-oriented worldview, which is not to be confused with OOP (object-oriented programming) because it doesn't specify the methods of objects, since in the common case this is not info to be transferred between federates. This view does only define how a federate must communicate with other federates, while it doesn't consider the internal representation of each federate. So, a simulation object model (SOM) is built, which defines what kind of data federates have to exchange with each other. Furthermore, a meta-object model, the federation object model (FOM), collects all the classes defined by each participant to the federation in order to give a description of all shared information [56](Fig. 4.6).

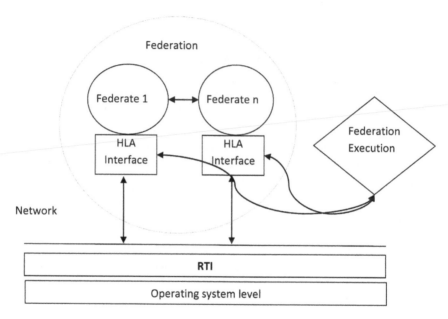

FIGURE 4.6   A distributed simulations under HLA.

The object models, then, describe the objects chosen to represent real world, their attributes and interactions and their level of detail.

Both the SOM and the FOM are based on the OMT, that is, on a series of tables that describe every aspect of each object. The OMT consists of the following 14 components:

1. Object model identification table
2. Object class structure table
3. Interaction class structure table
4. Attribute table
5. Parameter table
6. Dimension table
7. Time representation table
8. User-supplied tag table

9. Synchronization table
10. Transportation type table
11. Switches table
12. Data type tables
13. Notes table
14. FOM/SOM lexicon

HLA does not mandate the use of any particular FOM however several "reference FOMs" have been developed to promote a-priori interoperability. That is, in order to communicate, a set of federates must agree on a common FOM (among other things), and reference FOMs provide constructed FOMs that are supported by a wide variety of tools and federates [57].

Reference FOMs can be used or can be extended to add new simulation concepts that are specific to a particular federation or simulation domain. The RPR FOM (real-time platform-level reference FOM) is a reference FOM that defines HLA classes, attributes and parameters that are appropriate for real-time, platform-level simulations [58].

4.8 OPTIMIZATION

Its objective is to select the best possible decision for a given set of circumstances without having to enumerate all of the possibilities and involves maximization or minimization as desired. In optimization decision variables are variables in the model, which you have control over. Objective function is a function (mathematical model) that quantifies the quality of a solution in an optimization problem. Constraints must be considered, conditions that a solution to an optimization problem must satisfy and restrict decision variables are determined by defining relationships among them. It must be found the values of the decision variables that maximize (minimize) the objective function value, while staying within the constraints. The objective function and all constraints are linear functions (no squared terms, trigonometric functions, ratios of variables) of the decision variables [59, 60].

4.9 SIMULATION MODEL DEVELOPMENT

There are 11 steps which can expand the model simulation (Fig. 4.7):

Step 1. Identify Problem:
- Enumerate problems with an existing system
- Produce requirements for a proposed system

Step 2. a)   Formulate Problem:
- Define overall objectives of the study and specific issues to be addressed
- Define performance measures

b) Quantitative criteria on the basis of which different system configurations will be evaluated and compared
- Develop a set of working assumptions that will form the basis for model development
- Model boundary and scope (width of model)
- Determines what is in the model and what is out
- Level of detail (depth of model)
- Specifies how in-depth one component or entity is modeled
- Determined by the questions being asked and data availability
- Decide the time frame of the study
- Used for one-time or over a period of time on a regular basis

Step 3. a) Collect and Process Real System Data:
- Collect data on system specifications, input variables, performance of the existing system, etc.
- Identify sources of randomness (stochastic input variables) in the system
- Select an appropriate input probability distribution for each stochastic input variable and estimate corresponding parameters

b) Standard distributions
c) Empirical distributions
d) Software packages for distribution fitting

Step 4. a) Formulate and Develop a Model:
- Develop schematics and network diagrams of the system

b) How do entities flow through the system:
- ·Translate conceptual models to simulation software acceptable form
- ·Verify that the simulation model executes as intended

c) Build the model right (low-level checking):
- ·Traces
- ·Vary input parameters over their acceptable ranges and check the output

Step 5. a) Validate Model:

- Check whether the model satisfies or fits the intended usage of system (high-level checking)

b) Build the right model:

- Compare the model's performance under known conditions with the performance of the real system
- Perform statistical inference tests and get the model examined by system experts
- Assess the confidence that the end user places on the model and address problems if any

Step 6. Document Model for Future Use:

- Objectives, assumptions, inputs, outputs, etc.

Step 7. a) Select Appropriate Experimental Design:

- Performance measures
- Input parameters to be varied

b)   Ranges and legitimate combinations

c)   Document experiment design

Step 8. Establish Experimental Conditions for Runs:

- Whether the system is stationary (performance measure does not change over time) or nonstationary (performance measure changes over time)
- Whether a terminating or a nonterminating simulation run is appropriate
- Starting condition
- Length of warm-up period
- Model run length
- Number of statistical replications

Step 9. Perform Simulation Runs

Step 10. a) Analyze Data and Present Results:

- Statistics of the performance measure for each configuration of the model

b) Mean standard deviation, range, confidence intervals, etc.

- Graphical displays of output data

c) Histograms scatter plot, etc.

- Document results and conclusions

Step 11. Recommend Further Courses of Actions:

- ·   Other performance measures
- Further experiments to increase the precision and reduce the bias of estimators

- Sensitivity analysis
- How sensitive the behavior of the model is to changes of model parameters

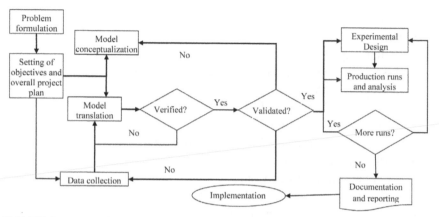

FIGURE 4.7   Steps in a simulation study.

4.10 SIMULATION LANGUAGES

SLAM introduced in 1979 by Pritsker and Pegden. SIMAN introduced in 1982 by Pegden, first language to run both a mainframe as well as a microcomputer. In primary computers, accessibility and interactions were limited. GASP IV introduced by Pritsker, Triggered a wave of diverse applications, which is significant in the evaluation of simulation.

Many programming systems have been developed, incorporating simulation languages. Some of them are general-purpose in nature, while others are designed for specific types of systems. FORTRAN, ALGOL, and PL/1 are examples of general-purpose languages, while GPSS, SIMSCRIPT, and SIMULA are examples of special simulation languages. Programming can be done in general purpose languages such as Java, simulation languages like SIMAN or use simulation packages, Arena [61, 62].

Four choices are existed: simulation language [63], general-purpose language [64], extension of general purpose [65], simulation package [66]. Simulation language is built in facilities for time steps, event scheduling, data collection, reporting. General-purpose is known to developer, available on more systems, flexible. The major difference is the cost tradeoff.

Simulation language requires startup time to learn, while general purpose may require more time to add simulation flexibility. Recommendation may be for all analysts to learn one simulation language so understand those "costs" and can compare. Extension of general-purpose is collection of routines and tasks commonly used. Often, base language with extra libraries that can be called and Simulation packages allow definition of model in interactive fashion. Get results in one day. Tradeoff is in flexibility, where packages can only do what developer envisioned, but if that is what is needed then is quicker to do so.

Now some of advantages and disadvantages of common languages are listed as:

a) General Simulation Languages: Arena, Extend, GPSS, SIMSCRIPT, SIMULINK (Matlab), etc.

- *Advantages:* Standardized features in modeling, shorter development cycle for each model, Very readable code
- *Disadvantages:* Higher software cost (up-front), Additional training required, limited portability

b) Special Purpose Simulation Packages: Manufacturing (Auto Mod, FACTOR/AIM, etc.), Communications network (COMNET III, NETWORK II.5, etc.), Business (BP$IM, Process Model, etc.), Health care (Med Model).

- *Advantages: Very* quick development of complex models, short learning cycle, little programming
- *Disadvantages:* High cost of software, limited scope of applicability, limited flexibility [67, 68].

4.11   SIMULATION CHECKLIST

You should check the simulation process in each step as:

a) Checks before developing simulation
- Is the goal properly specified?
- Is detail in model appropriate for goal?
- Does the team include the right mix (leader, modeling, programming, background)?
- · Has sufficient time been planned?

b) Checks during simulation development
- Is random number random?

- Is model reviewed regularly?
- Is model documented?

c) Checks after simulation is running
- Is simulation length appropriate?
- Are initial transients removed?
- Has the model been verified?
- Has the model been validated?
- Are there any surprising results? If yes, have they been validated?

4.12   COMMON MISTAKES IN SIMULATION

4.12.1   INAPPROPRIATE LEVEL OF DETAIL

Level of detail often potentially unlimited but more detail requires more time to develop and often to run. It can be introduced more bugs, making more inaccurate not less. Often, more detailed viewed as "better" but may not be the case. So more detail requires more knowledge of input parameters and getting input parameters wrong may lead to more inaccuracy. Therefore, start with less detail, study sensitivities and introduce detail in high impact areas.

4.12.2   IMPROPER LANGUAGE

Choice of language can have significant impact on time to develop, special-purpose languages can make implementation, verification and analysis easier, $C^{++}Sim$, Java Sim, Sim Py(thon).

4.12.3   UNVERIFIED MODELS

Simulations generally need large computer programs unless special steps taken, bugs or errors.

Invalid models: Unless no errors occur, the model does not represent real system, you need to validate models by analytic, measurement or intuition.

4.12.4   IMPROPERLY HANDLED INITIAL CONDITIONS

Often, initial trajectory is not representative of steady state so can lead to inaccurate results. In this case, typically you want to discard, but need a method to do it so effectively.

4.12.5   TOO SHORT SIMULATION RUNS

Attempting to save time makes even more dependent upon initial conditions. Therefore, correct length depends upon the accuracy desired (confidence intervals).

4.12.6   POOR RANDOM NUMBER GENERATORS AND SEEDS

"Home grown" are often not random enough to make artifacts. So the best is to use well-known one and choose seeds that are different [69, 70].

4.13   BETTER UNDERSTANDING BY A TANGIBLE EXAMPLE

4.13.1   BASIC KINETIC EQUATION

In recrystallization and transformation, a new phase forms and grows. These new phases continue to grow until they meet each other and stop growing. This situation is called hard impingement and can be expressed by using the Avrami type equation [71]:

$$X = 1 - \exp\left(-kt^n\right), k = \frac{\pi \dot{N} \dot{G}^3}{3}, n = 4 \tag{1}$$

or the Johnson–Mehl equation [72]. In these equations, the concept of extended volume fraction is adopted. By using this concept, the hard impingement can be taken into consideration indirectly. The extended volume fraction is the sum of the volume fraction of all new phases without direct consideration of the hard impingement between new particles and is related to the actual volume fraction by

$$X = 1 - \exp(-X_e) \tag{2}$$

The general form of the equation was developed by scientists. A brief explanation is presented here. The nucleation sites of new phases would be grain boundaries, grain edges, and/or grain corners. In the case of grain boundary nucleation, the volume fraction of a new phase after some time can be expressed as follows. Cahn considered the situation illustrated in Fig. 4.8 and calculated the volume of the semicircle.

In his calculation, firstly, the area at the distance of from the nucleation site is calculated. The summation of this area for all nuclei gives the total extended area. From this value, the actual area can be calculated. The extended volume can be obtained by integrating the area for all distances. Finally, the actual volume fraction can be derived.

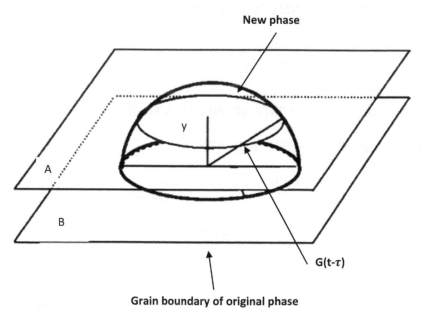

FIGURE 4.8   Schematic illustration of the situation of new phase at time , which nucleates at time at grain boundary .

The area of the section at a plane for a semicircle nucleated at a plane is considered. The radius r at time t can be expressed as:

$$r = \left[\dot{G}^2 (t - \tau)^2 - y^2 \right]^{1/2} \quad \text{For} \quad y < G(t - \tau) \tag{3}$$
$$r = o \qquad \qquad \text{For} \quad y > G(t - \tau)$$

In this calculation, the growth rate is assumed to be constant. From this radius, the extended area fraction for the new phases nucleated at time between and can be achieved as:

$$dY_e = \pi I_S d\tau \left[\dot{G}^2 (t - \tau)^2 - y^2 \right]^{1/2} \quad \text{For} \quad y < \dot{G}(t - \tau) \tag{4}$$
$$dY_e = o \qquad \qquad \text{For} \quad y > \dot{G}(t - \tau)$$

By integrating for the time from 0 to , the extended area fraction at the plane at time can be showed as:

$$Y_e = \int_0^t dY_e = \pi I_S \int_0^{t - y/\dot{G}} \left[\dot{G}^2 (t - \tau)^2 - y^2 \right] d\tau \tag{5}$$

By exchanging with , this equation changes to:

$$Y_e = \pi I_S \dot{G}^2 t^3 \left[\frac{1 - x^3}{3} - x^2 (1 - x) \right] \quad \text{For x<1} \tag{6}$$
$$Y_e = 0 \quad \text{For x>1}$$

The actual area fraction of new phases at plane , can be calculated by using :

$$Y = 1 - \exp(-Y_e) \tag{7}$$

The integration of for from to infinity gives the volume of new phases nucleated at unit area of plane , as:

$$V_0 = 2 \int_0^\infty Y dy = 2 \dot{G} t \int_0^1 \left[1 - \exp \exp \left\{ -\pi I_S \dot{G}^2 t^3 \left(\frac{1 - x^3}{3} - x^2 (1 - x) \right) \right\} \right] dx \tag{8}$$

Multiplying by the area of nucleation site, the extended volume fraction is obtained as:

$$X_e = SV_0 = b_s^{-\frac{1}{3}} f_s(a_s)$$

(9)

where

$$a_s = (I_S \dot{G}^2)^{\frac{1}{3}} t, b_s = \frac{I_S}{8S^3 \dot{G}} = \frac{\dot{N}}{8S^4 \dot{G}}$$

$$f_s(a_s) = a_s \int_0^1 \left[1 - \exp\exp\left\{ -\pi a_s^3 (\frac{1-x^3}{3} - x^2(1-x)) \right\} \right] dx$$

(10)

So the actual volume fraction can be expressed as:

$$X = 1 - \exp(-b_s^{-\frac{1}{3}} f_s(a_s))$$

(11)

From this equation, two extreme cases can be considered. One is the case where is very small and the other is extremely large. For these two cases, the equation becomes:

$$X = 1 - \exp\left(-\frac{\pi}{3\dot{N}\dot{G}^3 t^4} \right) a_s \ll 1$$

(12)

$$X = 1 - \left(-2S\dot{G}t \right) a_s \gg 1$$

(13)

The Eq. (12) is the same as the one obtained for the case of random nucleation sites by Johnson-Mehl. This equation implies that the increase in the volume of new phases is caused by nucleation and growth. On the other hand, Eq. (13) does not include nucleation rate and it implies that the nucleation sites are covered by new phases and the increase in the volume is dependent only on the growth of new phases. This situation is referred to as site saturation [73].

Cahn did this type of formulation for the cases of grain edge and grain corner nucleations. Table 4.2 shows all the extreme cases. For all cases, the increase of the volume of new phases for the case of small conforms to the case of nucleation and growth and site saturation for the case of large. The value of increases when the nucleation rate is small when compared to the growth rate. The early stage of reaction corresponds to small and the latter stage corresponds to large. From Table 4.2, we can recognize that the exponent of time depends on the mode of reaction and the type of nucleation site for the case of site saturation. The equations in Table 4.2 can be used for calculating actual reactions such as transformation and recrystallization by introducing fitting parameters [74].

TABLE 4.2 The Kinetic Equations Depending on the Modes and the Nucleation Sites of Reaction in Accordance with Cahn's Treatment

Nucleation site	Nucleation and growth	Site saturation
Grain boundary	$X = 1 - \exp\exp\left(-\dfrac{\pi}{3N_s G^3 t^4}\right)$	$X = 1 - \left(-2SGt\right)$
Grain edge		$X = 1 - \exp\left(-\pi LG^2 t^2\right)$
Grain corner		$X = 1 - \exp\left(-\left(\dfrac{4\pi}{3}\right)CG^3 t^3\right)$

4.13.2 UTILIZATION OF THERMODYNAMICS OF THE CALCULATION OF TRANSFORMATION AND RECITATION KINETICS

As transformation and precipitation kinetics are mostly related to phase equilibrium, thermodynamics can be used for their calculation. In this section, the method for using thermodynamics for the calculation will be explained.

For the consideration of kinetics, the Gibbs free-energy–composition diagram is much more useful and should be the basis. Figure 4.9 shows the Gibbs free-energy-composition diagram for austenite and ferrite in steels. Chemical composition at the phase interface between ferrite and austenite is obtained from the common tangent for free-energy curves of ferrite and austenite. The common tangent can be calculated under the condition that chemical potentials of all chemical elements in ferrite are equal to those in austenite. This condition is showed as:

$$\mu_i^\alpha = \mu_i^\gamma \tag{14}$$

In Fig. 4.9, the driving force for transformation from austenite to ferrite is indicated as well. It can be calculated by:

$$\Delta G_m = \sum x_i^\alpha (\mu_i^\gamma - \mu_i^\alpha) \tag{15}$$

These values are necessary for the calculation of moving rate of the interface during transformation and precipitation. The Zener–Hillert equation, which represents the growth rate of ferrite into austenite, is expressed as:

$$\dot{G} = \frac{1}{2r} D \frac{C_{\alpha\gamma} - C_\gamma}{C_\gamma - C_\alpha} \tag{16}$$

where , and is the carbon content in austenite, ferrite at interface and in ferrite apart from interface, respectively. The carbon content at the interface can be calculated from the common tangent between two phases as shown in Fig. 4.9. There is the other type of expression of moving rate of interface which is expressed as:

$$v = \frac{M}{V_m} \Delta G_m \tag{17}$$

The driving force in this equation can be calculated for multi component system by Eq. (15). This calculation makes it possible to consider the effect of alloying elements other than the pinning effect and the solute-drag effect [75]. Recently, some commercial software for the thermodynamic calculation has been used for this type of calculation [76].

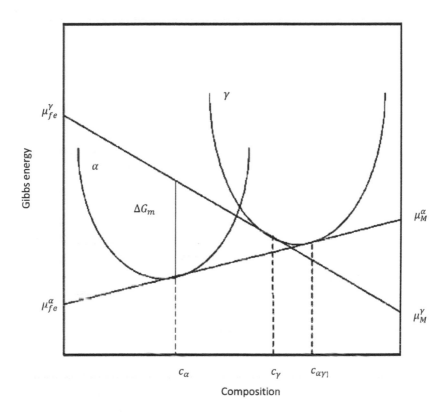

FIGURE 4.9   Gibbs free energy vs. chemical composition diagrams.

4.13.3 THE CONCEPT OF THE MODEL

The overall model for predicting mechanical properties of hot-rolled steels consists of several basic models: the initial state model for austenite grain size before hot-rolling, the hot-deformation model for austenitic micro structural evolution during and after hot-rolling, the transformation model for transformation during cooling subsequent to hot-rolling, and the relation between mechanical properties and microstructure of steels. In the case where steels include alloying elements, which form precipitates, the model for precipitation is essential [77].

4.13.3.1   INITIAL STATE MODEL

In this model, austenite grain sizes after slab reheating, namely before hot deformation, are calculated from the slab-reheating condition. In steels consisting of ferrite and pearlite at room temperature, austenite is formed between pearlite and ferrite and it grows into ferrite according to decomposition of pearlite. After all the microstructures become austenite, the grain growth of austenite takes place. We should formulate these metallurgical phenomena to predict austenite grain size after slab reheating. In hot-strip mill, however, the effect of initial austenite grain size on the final austenite grain size after multipass hot deformation is small. This can be due to the high total reduction in thickness by several hot-rolling steps in which the recrystallization and grain growth are repeated and the size of austenite grain becomes fine. This means that the high accuracy is not required for the prediction of the initial austenite grain size in a hot-strip mill. From this point of view, the next equation can be applied:

$$d_\gamma = \exp\left\{1.61\ln\ln\left(K + \sqrt{K^2 + 1}\right) + 5\right\} K = (T - 1413)/100 \tag{18}$$

On the other hand, the initial austenite grain size affects the final austenite grain size in the case of plate rolling because the total thickness reduction is relatively small compared to hot-strip rolling. In this case, the high accuracy of the prediction may be needed and the model that is suitable for this case has been investigated. Three steps are considered in this model: (i) the growth of austenite between cementite and ferrite according to the dissolution of cementite, (ii) the growth of austenite into ferrite at two-phase region, and (iii) the growth of austenite in the single-phase region. The pinning effect by fine precipitates on grain growth and that of Ostwald ripening of precipitates on the grain growth of austenite are taken into consideration. This model is briefly explained in the following paragraphs.

The growth of austenite due to the dissolution of cementites can be presented as:

$$\frac{d(d_\gamma)}{dt} = \frac{D_c^\gamma}{d_\gamma} \frac{C_{\theta\gamma} - C_{\gamma\alpha}}{C_{\gamma\alpha} - C_\alpha} \tag{19}$$

are the C content in austenite at phase interface and phase interface, respectively. In the two-phase region, the austenite grain size depends on the volume fraction of Austenite , which changes according to temperature. This situation is showed as:

$$d_\gamma = \left(\frac{3X_\gamma}{4\pi n_0}\right)^{1/3} \tag{20}$$

Grain growth happens in the austenite single-phase region. For grain growth, it is important to consider three cases: without precipitates, with precipitates, and with precipitates growing due to the Ostwald ripening. There are equations, which are formulated to theoretically correspond to these three cases. They are summarized by Nishizawa. The equation for the normal grain growth is expressed as:

$$d_\gamma^2 - d_{\gamma 0}^2 = k_2 t \tag{21}$$

where is the factor related to the diffusion coefficient inside the interface, the interfacial energy, and the mobility of the interface. With the pinning effect by precipitates, the growth rate becomes:

$$\frac{d(d_\gamma)}{dt} = M\left(\frac{2\sigma V}{R} - \Delta G_{pin}\right)\Delta G_{pin} = \frac{3\sigma Vf}{2r} \tag{22}$$

When precipitates grow according to the Ostwald ripening, the average size of precipitates used in the Eq. (22) is obtained:

$$r^3 - r_0^3 = k_3 t \tag{23}$$

where is the factor related to temperature, interfacial energy and the diffusion coefficient of an alloying element controlling the Ostwald ripening of precipitates. By this calculation method, it is possible to predict the growth of austenite grain during heating when precipitates exist in austenite.

4.13.3.2 HOT-DEFORMATION MODEL

The hot-deformation model is required to predict the austenitic microstructure before transformation through recovery, recrystallization, and grain growth in austenitic phase region during and after multipass hot deformation.

Sellars and Whiteman [78] made the first attempt on this issue and then several researchers [79, 80] developed models to calculate recovery, recrystallization, and grain growth. These models are basically similar to each other. In some models, dynamic recovery and dynamic recrystallization are taken into consideration. The dynamic recovery and recrystallization are likely to occur when the reduction is high for single-pass rolling or strain is accumulated due tomultipass rolling. They should be taken into consideration in finishing rolling stands of a hot-strip mill because, the interpass time might be less than 1 sec and the accumulation of strain might take place. Here, the hot-deformation model will be explained based on the model developed by Senuma et al. [79].

In this model, dynamic recovery and recrystallization, static recovery and recrystallization, and grain growth after recrystallization are calculated as shown in Fig. 4.10. The critical strain, at which dynamic recrystallization occurs is generally dependent upon strain rate, temperature, and the size of austenite grains. The effect of strain rate on is remarkable at low strain rate region [81].

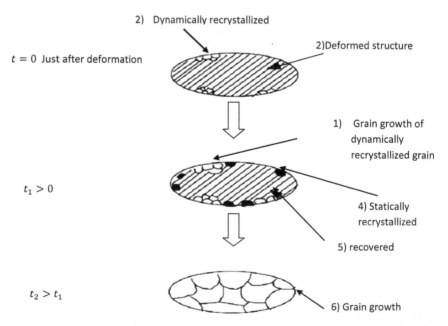

FIGURE 4.10 Schematic illustrations of micro structural are changes due to hot deformation.

One of the controversial issues had been whether the dynamic recrystallization took place or not when the strain rate is high such as that in a hot-strip mill. Researchers showed that it takes place and the effect of strain rate on is small at a high strain rate. The fraction dynamically recrystallized can be expressed based on the Avrami type equation as:

$$X_{dyn} = 1 - \exp\exp\left(-0.693\left(\frac{\epsilon - \epsilon_0}{\epsilon_{0.5}}\right)^2\right) \quad (24)$$

where is the strain at which the fraction dynamically re crystallized reaches 50%. On the other hand, the fraction statically recrystallized can be expressed as:

$$X_{dyn} = 1 - \exp\exp\left(-0.693\left(\frac{t - t_0}{t_{0.5}}\right)^2\right) \quad (25)$$

where $t_{0.5}$ is the time when the fraction statically re crystallized reaches 50% and t_0 is the starting time of static recrystallization.

The growth of grains re crystallized dynamically after hot deformation is much faster than normal grain growth in which grains grow according to square of time. This rapid growth was treated with different equations [79, 80]. The reason why this rapid growth takes place might be caused by the increase of the driving force for grain growth due to high dislocation density [79], the change in the grain boundary mobility or the annihilation of the small size grains at the initial stage. In the case of multipass deformation, the strain might not be reduced completely at the following deformation due to the insufficient time interval and the effect of accumulated strain on the recovery and recrystallization should be taken into consideration. This effect is remarkable for a hot-strip mill because of the short interpass time and for steels containing alloying elements, which retard the recovery and recrystallization. This effect can be formulated by using the change in the residual strain [82] or the dislocation density [79, 80]. In the modeling process, the accumulated strain is calculated from the average dislocation density which is obtained by calculating the changes in the dislocation density in the region dynamically recovered, and in the region recrystallized dynamically, according to time independently.

This method makes it possible to calculate the changes in grain size and dislocation density [83]. This model can be applied to the prediction of the resistance to hot deformation as well and it can contribute to the improvement of the accuracy in thickness. In this method, the average values concerning the grain size and the accumulated dislocation density are used taking the fraction re crystallized into consideration. This averaging can be applied to the hot-strip mill because the total thickness reduction is large enough to recrystallize their microstructure. In the case of plate rolling, the use of the average values is unsuitable because the reduction at each pass is small and the total thickness reduction is not enough to re crystallize the microstructure of steels. The model applicable to this case has been developed by dividing the microstructure into several groups [84, 85].

KEYWORDS

- **Computational models**
- **Microstructure**
- **Modeling and Simulation Principles**
- **Optimization**
- **Physical Systems for Particle Models**
- **Static re-crystallization**

CHAPTER 5

MECHANIC QUANTUM AND THERMODYNAMIC IN NANOELEMENTS

CONTENTS

5.1   INTRODUCTION

A revolution is occurring in science and technology, based on the recently developed ability to measure, manipulate and organize matter on the nanoscale 1 to 100 billionths of a meter. At this level everything is attenuated to fundamental interactions between atoms and molecules. Therefore, physics, chemistry, biology, materials science, and engineering converge toward the same principles and tools. A nanoelement compares to a basketball, like a basketball to the size of the earth. The aim of nonscientists is to manipulate and control the infinitesimal particles to create novel structures with unique properties. The science of atoms and simple molecules and the science of matter from microstructures to larger scales are generally established, in parallel. The remaining size related challenge is at the nanoscale where the fundamental properties of materials are determined and can be engineered. A revolution has been occurring in science and technology, based on the ability to measure, manipulate and organize matter on this scale. These properties are incorporated into useful and functional devices. Therefore, nanoscience will be transformed into nanotechnology. Through a basic understanding of ways to control and manipulate matter at the nanometer scale and through the incorporation of nanostructures and nano-processes into technological innovations, nanotechnology will provide the capacity to create affordable products with dramatically improved performance. Nanotechnology involves the ability to manipulate, measure, and model physical, chemical, and biological systems at nanometer dimensions, in order to exploit nanoscale phenomena [1].

Novel properties in biological, chemical, and physical systems can be approximately obtained at dimensions between 1 nm to 100 nm. These properties can differ in fundamental ways from the properties of individual atoms and molecules and those of bulk materials [1]. Nowadays, advances in nanoscience and nanotechnology indicate to have major implications for health, wealth, and peace. Knowledge in this field due to fundamental scientific advances, will lead to dramatic changes in the ways that materials, devices, and systems are understood and created. Nanoscience will redirect the scientific approach toward more generic and inter disciplinary research [1, 2].

Nanoelement categories consist of atom clusters/assemblies or structures possessing at least one dimension between 1 and 100 nm, containing 10^3–10^9 atoms with masses of 10^4–10^{10} Daltons. Nanoelements are homogenous, uniform nanoparticles exhibiting well-defined (a) sizes, (b) shapes, (c) surface chemistries, and(d) flexibilities (i.e., polarizability). Typical nanoelement categories exhibit certain nanoscale atom mimicry features such as (a) core–shell architectures, (b) predominately (0-D) zero dimensionality (i.e., 1-D in some cases), (c) react and behave as discrete, quantized modules in their manifestation of nanoscale physicochemical properties, and(d)display discrete valencies, stoichiometries, and mass combining ratios as a consequence of active atoms or reactive/passive functional groups presented in the outer valence shells of their core-shell architectures. Nanoelements must be accessible by synthesis or fractionation/separation methodologies with typical mono dispersities (90%) (i.e., uniformity) [3] as a function of mass, size, shape, and valency. Wilcoxon et al. have shown that hard nanoparticle Au nano-clusters are as mono disperse as 99.9% pure (C) [3]. Soft nanoparticle dendrimers are routinely produced as high as generation = 6–8 with poly dispersities ranging from 1.011 to 1.201 [4–6]. Nanoelement categories must be robust enough to allow reproducible analytical measurements to confirmed size, mass, shape, surface chemistries, and flexibilities/polarizability parameters under reasonable experimental conditions (Table 5.1).

TABLE 5.1 The Importance of Scales

Length(m)	1	10^{-1}	10^{-2}	10^{-3}	10^{-4}	10^{-5}	10^{-6}	10^{-7}	10^{-8}	10^{-9}
Physical Laws	Macroscopic					Mesoscopic		Microscopic		
Science	Physics (Classical) Biology (Convectional) Engineering (Almost All)				Physics (Solid state) Biology(Microbio) Material Science			Physics (Molecular) Biology (Molecular) Chemistary		
Technology	Bulk Technology					Microtechnology			Nanotech	
How to see	Human eyes					Optical Microscope		Electron Microscope		SPM
Simulation Approach	Macroscale					Mesoscale		Nanoscale		QM
Successful Model	← Empirical					First Principles		→		

5.2   THE RELATIONSHIP BETWEEN NANOSCIENCE AND MECHANIC QUANTUM

The nanoscale is not just another step toward miniaturization, but a qualitatively new scale. At these sizes, nano-systems can exhibit interesting and useful physical behaviors based on quantum phenomena. The new behavior is dominated by quantum mechanics, material confinement in small structures, large interfacial volume fraction, and other unique properties, phenomena and processes. Atom (element)-based chemistry discipline" before the advent of quantum mechanics and electronic theory, Dalton's atom/molecular theory is:

1.   Each element consists of pi co scale particles called atoms.
2.   The atoms of a given element are identical; the atoms of different elements are different in some fundamental way(s).
3.   Chemical compounds are formed when atoms of different elements combine with each other. A given compound always has the same relative number in types of atoms.
4.   Chemical reactions involve reorganization of atoms (i.e., changes in the way they are bound).

Critical parameters that allowed this important progress evolved around discrete, reproducible features exhibited by each atomic element such as well-defined (a) atomic masses, (b) reactivates, (c) valency, (d) stoichiometries, (e) mass-combining ratios, and (f) bonding directionalities. This intrinsic elemental properties inherents in all atom-based elemental structures.

Isaac Newton created, more than 300 years ago, classical mechanics by finding the laws of motion for solids and of gravitation between masses. This theory was so successful for the deterministic description of motions. At the beginning of the twentieth century, then, experimental results accumulated which contributed essentially to the emergence of a new physics, quantum physics. Also known as quantum or wave mechanics, this branch of physics was created by Max Planck who showed that the exchange of energy between matter and radiation occurred in discontinuous quantities (quanta). The quantum mechanics is presented as one of the most important and successful theories to solve physical problems. This is totally in the sense of most physicists, who applied, until the 1970 s of the twentieth century, in a first quantum revolution quantum mechanics with overwhelming success not only to atom and particle physics but also to nearly all other science branches as chemistry, solid state physics, biology or astrophysics. Because of the success in answering essential questions in these fields fundamental open problems concerning the theory itself were approached only in rare cases. This situation has changed since the last decade of the twentieth century [7].

The "second quantum revolution" as this continuing further development of quantum physical thinking is called by Alain Aspect, one of the pioneers in this fields one expects a deeper understanding of quantum physics itself but also applications in engineering. There is already the term "quantum engineering" which describes scientific activities to apply particle wave duality or entanglement for practical purposes, for example, nano-machines, quantum computers etc. [8, 9] (Fig. 5.1).

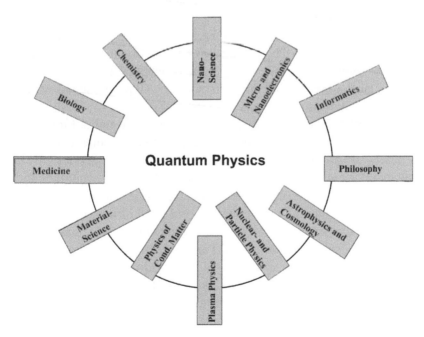

FIGURE 5.1   The relationship between important science branches and the fields of quantum physics.

The nano-world is part of our world, but in order to understand this, concepts other than the normal ones, such as force, speed, weight, etc., must be taken into consideration. The nano-world is subject to the laws of quantum physics, yet evolution has conditioned us to adapt to this ever-changing world. This observation has led to further investigate theories based on the laws of physics that deal with macroscopic phenomena. In the macro world, sizes are continuous; however, this is not the case in the nano-world. When we investigate and try to understand what is happening on this scale, the way we look at things must be changed. New concepts of quantum physics can only come directly from our surroundings. However, our world is fundamentally quantum. Our common sense in this world has no value in the nano-world [10].

Quantum physics gives a completely different version of the world on the nano-metric scale than that given by traditional physics. A molecule is described by a cloud of probability with the presence of electrons at discrete energy levels. This can only be represented as a simulation. All measurable sizes are subject to the laws of quantum physics, which condition every organism in the world, from the atom to the different states of matter. The nano-world must therefore be addressed with quantum concepts. Chemistry is quantum. The chemistry of living organisms is quantum. Is the functioning of our brain closer to the concept of a quantum computer or to the most sophisticated microprocessors? All properties of matter are explicable only by quantum physics [10].

Traditional physics, which is certainly efficient and sufficient in the macroscopic domain only deals with large objects (remember that there are nearly 10^{23} atoms per cm^3 in a solid), while quantum physics only deals with small discrete objects. However, the evolution of techniques and the use of larger and larger objects stemming from scientific discoveries make us aware of the quantum nature of the world in all its domains. Everything starts with the atom, the building block of the nano-world, and also of our world. In mechanic quantum view, Particles can behave like waves. This property, particularly for electrons, is used in different investigation. On the other hand, waves can also act like particles: the photoelectric effect shows the corpuscular properties of light [10].

5.2.1   THE WAVE FUNCTION AND ITS INTERPRETATION

It has been proven that light waves propagating in space as well as atomic and subatomic particles as electrons moving from one to another spot have one thing in common: Their propagation obeys the laws of wave expansion.

In fact, it can be said that everything, matter and energy fields, are simultaneously wave and particle. The correspondence between particle and wave can be expressed by the following relations:

$$E = \frac{1}{2}mv^2 = \hbar\omega \tag{1}$$

$$p = mv = \hbar k = \hbar\frac{2\pi}{\lambda}\frac{k}{|k|} \tag{2}$$

The propagation of a particle, for example, of an electron is described by a wave function. In the simplest case of motion along a straight line a plane wave describes the propagation of the particle, where wave vector and frequency are connected to the particle:

$$\psi(r,t) = ce^{i(k.r-\omega t)} \tag{3}$$

The wave function is a quantity, which is analogous to the wave amplitude of a light fields. Its absolute square is identified with an observed intensity after collecting a huge number of electrons on a screen. In particular, the interference pattern in a double slit experiment with electrons is obtained by superimposing two waves originating from two slits at the positions on a remote screen. At a long distance from the source both spherical and cylinder waves (circular holes or slits) can be approximated by plane waves. At the observation point on the remote screen, the superposition of the two wave functions thus yields (Fig. 5.2).

$$\psi = \psi_1 + \psi_2 \qquad with \qquad \psi_i = ce^{i[k.(r-r_i)-\omega t]} \tag{4}$$

The intensity can be shown as:

$$I = |\psi(r,t)|^2 = |\psi_1|^2 + |\psi_2|^2 + 2c^2\cos k.(r_2 - r_1) \tag{5}$$

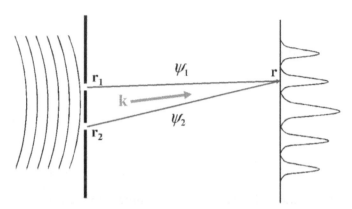

FIGURE 5.2   Scheme of double slits interferences of two particle waves.

The wave function is a statistical quantity, which describes only ensemble properties. The probability to find an electron in a volume element is proportional to the volume and, of course, to the probability that is:

$$dP \propto \left| \psi(r,t) \right|^2 d^3r \qquad (6)$$

The total probability to find the particle somewhere in the volume must be written as:

$$P(particle \quad in \quad V) = \int_v d^3r \left| \psi(r,t) \right|^2 = 1 \qquad (7)$$

The wave function as a probability density must be normalized, in the sense of above equation over the volume of the whole system considered. Depending on the particular problem the considered volume might be the whole universe.

5.2.2   WAVE PACKET AND PARTICLE VELOCITY

The energy-frequency relation [Eq. (1)], the connection between particle momentum and wave number [Eq. (2)] as well as the description of particle propagation by a wave function and its statistical interpretation [Eqs. (3), (4) and (6)], are the starting point for the formal description of the particle-wave duality.

For a spatially extended wave in the extreme limit, over the whole space-the velocity of a particle cannot be described. The term velocity contains inherently the movement of a particle, an entity, which is more or less limited in its spatial extension.

A particle with a spatial extension in one dimension might be described in simple approximation by a wave function having Gaussian shape.

$$\psi(x) = \frac{1}{\sqrt{2\pi}} \int_{-\infty}^{\infty} a(k)e^{ikx}dk \tag{8}$$

$$a(k) = \frac{1}{\sqrt{2\pi}} \int_{-\infty}^{\infty} \psi(x)e^{ikx}dx \tag{9}$$

$$\psi(x) = \left[2\pi(\Delta x)^2\right]^{-\frac{1}{4}} \exp\left(-\frac{x^2}{4(\Delta x)^2}\right) \tag{10}$$

$$a(k) = \frac{1}{\sqrt{2\pi}} \int_{-\infty}^{\infty} \left[2\pi(\Delta x)^2\right]^{-\frac{1}{4}} \exp\left(-\frac{x^2}{4(\Delta x)^2}\right) \exp(-ikx)$$

$$= \frac{1}{\sqrt[4]{(2\pi)^3(\Delta x)^2}} \int_{-\infty}^{\infty} \exp\left(-\frac{x^2}{4(\Delta x)^2}\right) \exp(-ikx)dx \tag{11}$$

Finally it can be obtained:

$$a(k) = \left(\frac{2}{\pi}\right)^{\frac{1}{4}} (\Delta x)^{\frac{1}{2}} \exp\left[-(\Delta x)^2 k^2\right] = \left[\frac{4(\Delta x)^2}{2\pi}\right]^{\frac{1}{4}} \exp\left[-(\Delta x)^2 k^2\right] \tag{12}$$

If Eq. (12) is compared with the common representation of a Gauss distribution as function of k with width Δk, it can be expressed as:

$$a(k) = \frac{1}{2\pi(\Delta k)^2} \exp\left[-\frac{k^2}{4(\Delta k)^2}\right] \tag{13}$$

The following relation between spatial width Δx of the wave packet and the spread or width of the corresponding wave vector distribution Δk can be obtained as:

$$\Delta k \Delta x = \frac{1}{2} \qquad (14)$$

5.2.3 THE UNCERTAINTY PRINCIPLE

From the representation of a particle by means of a wave packet, we conclude directly that the width Δk of the distribution of wave vectors a(k) which constitute the wave packet is inversely proportional to the spread, that is, the spatial extension of the wave packet (see Eq. (14)). For a Gaussian wave packet we quantitatively obtain the relation (Fig. 5.3).

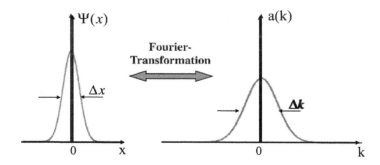

FIGURE 5.3   Gaussian wave packet with a spatial extension (full width at half maximum) and its Fourier representation in the wave number space.

Because of the general rules of Fourier transformation a relation similar to (see Eq. (14)) is always valid:

$$\Delta x . \Delta k \approx 1 \qquad (15)$$

This relation between the spatial width Δx of a wave packet and the spread Δk of its Fourier transform leads to an important, typically quantum physical phenomenon.

5.3   THE VISION FOR NANOMATERIALS TECHNOLOGY

Nanomaterials will deliver new functionality and options of material types. A diverse range of nano-material building blocks with well-defined properties and stable compositions will enable the design of nanomaterials that provide levels of functionality and performance which are not available in conventional materials.

Manufacturers will combine the benefits of traditional materials and nanomaterials to create new generations of nano-material-enhanced products that can be seamlessly integrated into complex systems. In some occasions, nanomaterials will serve as stand-alone devices, providing incomparable functionality. Nanomaterials show a prodigious opportunity for industry to introduce a host of new products that would energize the economy, solve major societal problems, revitalize existing industries, and create entirely new businesses. The race to research, develop, and commercialize nanomaterials is obviously global.

The nanoscience concept proposed the following: (a) creation of a nanomaterials roadmap focused solely on well defined (i.e., >90% mono disperse), (0-D) and (1-D) nanoscale materials; (b) these well defined ma terials were divided into hard and soft nanoparticles, broadly following compositional/architectural criteria for traditional inorganic and organic materials; (c) a preliminary table of hard and soft nanoelement categories consisting of six (6) hard matter and six (6) soft matter particles was proposed. Elemental category selections were based on "atom mimicry" features and the ability to chemically combine or self assemble like atoms; (d) these hard and soft nanoelement categories produce a wide range of stoichiometric nanostructures by chemical bonding or nonbonding assembly. An abundance of literature examples provides the basis for a combinatorial library of hard-hard, hard-soft and soft-soft nano-compounds, many of which have already been characterized and reported. However, many such predicted constructs remain to be synthesized and characterized; (e) based on the presumed conservation of critical module design parameters, many new emerging nano-periodic property patterns have been reported in the literature for both the hard and soft nanoelement categories and their compounds [11] (Fig. 5.4).

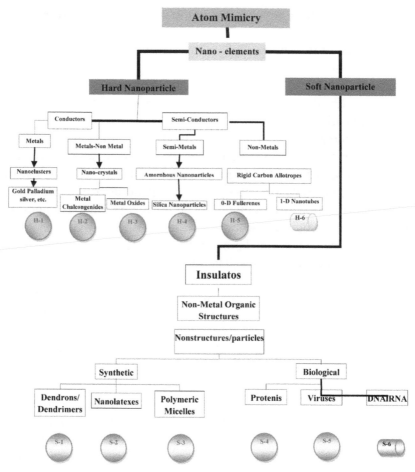

FIGURE 5.4   Nanomaterials classifications road map.

5.4   THE FUNDAMENTAL IMPORTANCE OF SIZE

Some of the technologies deal with systems on the micrometer range and not on the nanometer range (1–100 nm). In fact, distance scales used in science go to much smaller than nanometers and much larger than meters. All the experience in the macroscopic world suggests that matter is continuous. This, however, leads to a paradox because if matter were a continuum, it could be cut into smaller and smaller pieces without end. If one were able to keep cutting a piece of matter in two, each of those pieces into two, and so on ad infinitum, one could, at least in principle, cut it out of existence into pieces of nothing that could not be reassembled.

Nowadays it can be studied pieces of matter of smaller and smaller size right down to the atom. The important result is that the properties of the pieces start to change at sizes much bigger than a single atom. When the size of the material crosses into the nano-world, its fundamental properties start to change and become dependent on the size of the piece. It is an important issue to know how the behavior of a piece of material can become critically dependent on its size [12].

Nanomaterials have an increased surface-to-volume ratio compared to bulk materials. Beginning with the most clearly defined category, zero-dimensional nanomaterials are materials where in all the dimensions are measured within the nanoscale. On the other hand, 1-D nanomaterials differ from 0-D nanomaterials in that the former have one dimension that is outside the nanoscale. This difference in material dimensions leads to needle like-shaped nanomaterials. Two-dimensional nanomaterials are more difficult to classify. Three-dimensional nanomaterials, also known as bulk nanomaterials, are relatively difficult to classify. Nowadays it can be studied pieces of matter of smaller and smaller size right down to the atom. The important result is that the properties of the pieces start to change at sizes much bigger than a single atom. Top down approach refers to slicing or successive cutting of a bulk material to get nano-sized particle which enables to control the manufacture of smaller, more complex objects, as illustrated by micro and nano-electronics. Bottom up approach refers to the buildup of a material from the bottom: atom-by-atom, molecule-by-molecule or cluster-by-clusterwhich enables to control the manufacture of atoms and molecules, as illustrated by supra molecular chemistry. Both approaches play very important role in modern industry and most likely in nanotechnology as well. Directed and high rat self-assembly are the efficient methods for nanoelements production which classified into combined top-down and bottom-up nano-manufacturing.

5.5 THERMODYNAMICS AND STATISTICAL MECHANICS OF SMALL SYSTEMS

This section is about the important subjects of computational nanotechnology, namely thermodynamics and statistical mechanics and their applications in molecular systems to predict the properties and performances involving nanoscale structures. A scientific and technological revolution has begun in our ability to systematically organize and manipulate matter on a bottom-up fashion starting from atomic level as well as design tools, machinery and energy conversion devices in nanoscale towards the development of nanotechnology. There is also a parallel miniaturization activity to scale down large tools, machinery and energy conversion systems to micro and nanoscales towards the same goals [13, 14].

Principles of thermodynamics and statistical mechanics for macroscopic systems are well defined and mathematical relations between thermodynamic properties and molecular characteristics are derived. The objective here is to introduce the basics of the thermodynamics of small systems and introduce statistical mechanical techniques, which are applicable to small systems. This will help to link the foundation of molecular based study of matter and the basis for nanoscience and technology.

The subject of thermodynamics of small systems was first introduced by Hill [16, 17] in two volumes in 1963 and 1964 [15] to deal with chemical thermodynamics of mixtures, colloidal particles, polymers and macromolecules. Nano-thermodynamics, a term which is recently introduced in the literature by Hill [16, 17], is a revalidation of the original work of Hill mentioned above on thermodynamics of small systems.

5.5.1 THERMODYNAMIC SYSTEMS IN NANOSCALE

The definition of a thermodynamic system in nanoscale is the same as the macroscopic systems. In thermodynamics, a system is any region completely enclosed within a well-defined boundary. Everything outside the system is then defined as the surroundings. The boundary may be either rigid or movable. It can be impermeable or it can allow heat, work or mass to be transported through it. In any given situation a system may be defined in several ways.

The simplest system in nanoscale may be chosen as a single particle, like an atom or molecule, in a closed space with rigid boundaries. In the absence of chemical reactions, the only processes in which it can participate are transfers of kinetic or potential energy to or from the particle, from or to the walls. The state for this one-particle system is a set of coordinates in a multidimensional space indicating its position and its momentum in various vector directions.

The set of all the thermodynamic properties of a multiparticle system including its temperature, pressure, volume and internal energy is defined as the thermodynamic state of this system. An important aspect of the relationships between thermodynamic properties in a large, macroscopic and also known as extensive system is the question of how many different thermodynamic properties of a given system are independently variable. The number of these represents the smallest number of properties, which must be specified in order to completely determine the entire thermodynamic state of the system.

5.5.2 LAWS OF THERMODYNAMICS IN NANO-SYSTEMS

The application of thermodynamics of large and small systems to the prediction of changes in given properties of matter in relation to energy transfers across its boundaries is based on four fundamental axioms, the Zeroth First, Second, and Third Laws of thermodynamics. The question whether these four axioms are necessary and sufficient for all systems whether small or large, including nano-systems.

a) The Zeroth Law of thermodynamics consists of the establishment of an absolute temperature scale.

b) The First Law of thermodynamics as defined for macroscopic systems in which no nuclear reactions is taking place is simply the law of conservation of energy and conservation of mass. When, due to nuclear reactions, mass and energy are mutually interchangeable, conservation of mass and conservation of energy should be combined into a single conservation law.

$$dE = \delta Q_{in} + \delta W_{in} \tag{16}$$

Transfer of energy through work mode is a visible phenomenon in macroscopic systems. However, it is invisible in a nano-system, but it occurs as a result of the collective motion of an assembly of particles of the nano-system resulting in changes in energy levels of its constituting particles. Transfer of energy through heat mode is also an invisible phenomenon, which occurs in atomic and molecular level. It is caused by a change not of the energy levels but of the population of these levels.

a) Lord Kelvin originally proposed the Second Law of thermodynamics in the nineteenth century. He stated that heat always flows from hot to cold. Rudolph Claussius later stated that it was impossible to convert all the energy content of a system completely to work since some heat is always released to the surroundings. Kelvin and Claussius had macro systems in mind where fluctuations from average values are in significant in large time scales. According to the Second Law of thermodynamics for a closed (controlled mass) system we have [18].

$$dP_s = ds - \frac{\delta Q_{in}}{T_{ext}} \geq 0 \tag{17}$$

b) The Third Law of thermodynamics for large systems, also known as "the Nernst heat theorem," states that the absolute zero temperature are unattainable. Currently, the third law of thermodynamics is stated as a definition: the entropy of a perfect crystal of an element at the absolute zero of temperature is zero. This definition seems to be valid for the small systems as well as the large systems.

Recent developments in nanoscience and nanotechnology have caused a great deal of interest into the extension of thermodynamics and statistical mechanics to small systems consisting of countable particles below the thermodynamic limit. Hence, if we like to extend thermodynamics and statistical mechanics to small systems in order to remain on a firm basis we must go back to its founders and, like them establish new formalism of thermodynamics and statistical mechanics of small systems starting from the safe grounds of mechanics.

Structural characteristics in nanoscale systems are dynamic, not the static equilibrium of macroscopic phases. Coexistence of phases is expected to occur over bands of temperature and pressure, rather than along just sharp points. The pressure in a nano-system cannot be considered isotropic and must be generally treated as a tensor.

The Gibbs phase rule loses its meaning, and many phase-like forms may occur for nanoscale systems that are unobservable in the macroscopic counterparts of those systems [15, 19].

5.5.3 STATISTICAL MECHANICS OF NANO-SYSTEMS

The objective of statistical mechanics is generally to develop predictive tools for computation of properties and local structure of fluids, solids and phase transitions from the knowledge of the nature of molecules comprising the systems as well as intra and intermolecular interactions.

The accuracy of the predictive tools developed through statistical mechanics will depend on two factors. The accuracy of molecular and intermolecular properties and parameters available for the material in mind and the accuracy of the statistical mechanical theory used for such calculations.

In the case of nano-(small) scale there is little or no such data available and the molecular theories of matter in nanoscale are in their infancy. With the recent advent of tools to observes, study and measure the behavior of matter in nanoscale it is expected that in a near future experimental nanoscale data will become available.

Recent nanotechnology advances, both bottom-up and top-down approaches, have made it possible to envision complex and advanced systems, processes, reactors, storage tanks, machines and other moving systems which include matter in all possible phases and phase transitions. So in the next section several methods of manufacturing nanoscale will be reviewed.

5.6 NANOELEMENT MANUFACTURING

5.6.1 MANUFACTURING AT NANOSCALE DIMENSIONS

The physicochemical properties of nano-sized materials are really unprecedented, exquisite and sometimes even adjustable in contrast to the bulk phase. For instance, quantum confinement phenomena allow semiconductor nanoparticles to sustain a dilating of their band gap energy as the particle size becomes smaller. Thereby it causes the blue-shifts in the optical spectra and a change in their energy density from continuous to discrete energy levels as the transition moves from the bulk to the nanoscale quantum dot state [20–23]. In addition, interesting electrical properties including resonance tunneling and Coulomb blockade effects are observed with metallic and semiconducting nanoparticles, and endohedral fullerenes and carbon nanotubes can be processed to exhibit a tunable band gap of either metallic or semiconducting properties [24, 25]. These very different phenomena are mainly due to larger surface area-to-volume ratio at the nanoscale compared to the bulk. Thus, the surface forces become more important when the nano-sized materials exhibit unique optical or electrical properties. The surface (or molecular) forces can be generally categorized as electro- static, hydration (hydrophobic, and hydrophilic), Vander Waals, capillary forces, and direct chemical interactions [26, 27]. Based on these forces, the synthesis and processing techniques of these interesting nano-sized materials have been well established as capable of producing high-quality mono-disperse nano-crystals of numerous semiconducting and metallic materials, fullerenes of varying properties, single- and multiwall carbon nanotubes, conducting polymers, and other nano-sized systems [28]. The next key step in the application of these materials to device fabrication is undoubtedly the formation of subnanoelements into functional and desired nanostructures without mutual aggregation. To achieve the goal of innovative developments in the areas of microelectronic, optoelectronic and photonic devices with unique physical and chemical characteristics of the nano-sized materials, it maybe necessary to immobilize these materials on surfaces and/or assemble the min to an organized network [27].

Many significant advances in one-to three-dimensional arrangements in nanoscale have been achieved using the 'bottom-up' approach. Unlike typical top-down photolithographic approaches, the bottom-up process offers numerous attractive advantages, including the substantiation of molecular-scale feature sizes, the potential of three-dimensional assembly and an economical mass fabrication process [29]. Self-assembly is one of the few vital techniques available for controlling the orchestration of nanostructures via this bottom-up technology. The self-assembly process is defined as the autonomous organization of components into well-organized structures. It can be characterized by its numerous advantages such as cost-effective, versatile, facile, and the process seeks the thermodynamic minima of a system, resulting instable and robust structures [30]. As the description suggests, it is a process in which defects are not energetically favored, thus the degree of perfect organization is relatively high. As described earlier, there are various types of interaction forces by which the self-assembly of molecules and nanoparticles can be accomplished [31, 32].

5.6.2 OVERVIEW OF MOLECULAR SELF-ASSEMBLY

Molecular self-assembly is the assembly of molecules without guidance or management from an outside source. Self-assembly can happen spontaneously in nature, for example, in cells such as the self-assembly of the lipid bi-layer membrane. It usually results in an increase in internal organization of the system. Many biological systems use self-assembly to assemble various molecules and structures. Imitating these strategies and creating novel molecules with the ability to self-assemble into supra molecular assemblies is an important technique in nanotechnology [33–35].

In self-assembly, the final desired structure is 'encoded' in the shape and properties of the molecules that are used, as compared to traditional techniques, such as lithography, where the desired final structure must be carved out from a larger block of matter [36].

On a molecular scale, the accurate and controlled application of inter molecular forces can lead to new and previously unachievable nanostructures. This is why molecular self- assembly (MSA) is a highly topical and promising field of research in nanotechnology today. With many complex examples all around in nature, MSA is a widely perceived phenomenon that has yet to be completely understood. Bio molecular assemblies are sophisticated and often hard to isolate, making systematic and progressive analyzes of their fundamental science very difficult. What in fact are needed are simpler MSAs, the constituent molecules of which can be readily synthesized by chemists. These molecules would self-assemble into simpler constructs that can be easily assessed with current experimental techniques [37, 38].

Of the diverse approaches possible for Molecular Self-assembly, two strategies have received significant research attention, electrostatic Self-assembly (or layer- by-layer assembly) and "Self-assembled Mono layers (SAMs). Electrostatic self-assembly involves the alternate adsorption of anionic and cationic electrolytes onto a suitable substrate. Typically, only one of these is the active layer while the other enables the composite multilayered film to be bound by electrostatic attraction. The latter strategy of Self-assembled mono layers or SAMs based on constituent molecules, such as thiols and silanes [39, 40].

For SAMs, synthetic chemistry is used only to construct the basic building blocks (that is the constituent molecules), and weaker intermolecular bonds such as Vander Waals bonds are involved in arranging and binding the blocks together into a structure. This weak bonding makes solution, and hence reversible, processing of SAMs (and in general, MSAs) possible. Thus, solution processing and manufacturing of SAMs offer the enviable goal of mass production with the possibility of error correction at any stage of assembly. It is well recognized that, this method could prove to be the most cost-effective way for the semiconductor electronics industry to produce functional nano-devices such as nano-wires, nano-transistors, and nano-sensors in large numbers [41, 42].

5.6.3   NANO-SELF-ASSEMBLY INVESTIGATION

In the previous sections self-assembly was defined as assembly of its building units. All possible entities (atoms, molecules, colloidal particles) that can take part in this process are self-assembly building units. Building units for nanotechnology systems have more structural hierarchies. Nanotechnology systems can be built not only through self-assembly processes but through an external manipulation as well. All these efforts to create nanotechnology systems can be considered as the processes for assembling nano-technology systems. We will define this as a nano-assembly, which can be stated as a "thermodynamic, kinetic, or manipulative assembly of nano-assembly building units." Spontaneous assembly of nano-assembly building units will be a great route for building nanotechnology systems [43, 44].

However, assembling them, for example, using an atomic force microscope through a one-by-one type of operation with any type of nano-assembly building units will also be a great alternative for creating nanotechnology systems. Figure 5.5 (left-hand side) shows that nano-assembled systems are assembled from three basic nano-assembly building units. They are a self-assembly building unit, a fabrication-building unit, and a reactive building unit. As will be described in the next section with more details, the structures of all three basic nano-assembly-building units can be analyzed based on the concept of segmental analysis. In other words, the segmental analysis that was developed for self-assembly building units can be expanded for the two other types of building units. Figure 5.6 explains this. All three basic nano-assembly building units can be analyzed with the three fundamental and two additional segments. And all segments from the three basic nano-assembly-building units interact through the force balance with any possible combinations. The whole process resembles the self-assembly process. But it now occurs in a "quasi-three dimensional" way, which is to imply that there are three different types of building units instead of just one (as for self-assembly). The concept of force balance is directly applied not only between self-assembly building units, between fabrication building units, or between reactive building units, but between all three different types of building units as well. This gives us an important insight for the third part of this book that there can be great possibilities for building nano-assembled systems once the three basic building units are well identified and the relationships between them are well controlled. The roles of the five segments during the assemblies of nano-assembled systems are the same as for the assemblies of self-assembled systems.

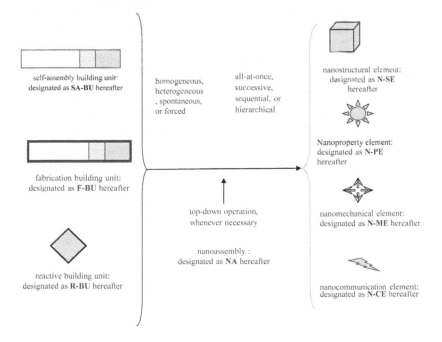

FIGURE 5.5   Three basic nano-assembly-building units construct the four nanoelements. Force balance between the nano-assembly building units plays a key role.

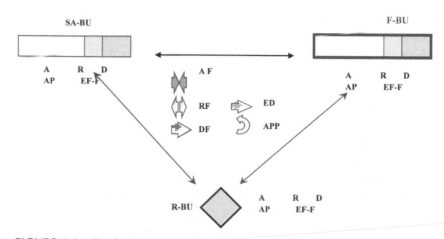

FIGURE 5.6   The fundamental and additional segments of self-assembly building unit (SA-BU), fabrication building unit (F-BU), and reactive building unit (R-BU) can interact through the force balance with any possible combinations [AF, RF, and DF represent attractive force between as, repulsive force between Rs, and directional force between Ds, respectively. A, R, D, AP, and EF-F refers to attractive, repulsive, directional, asymmetric packing, and external force–specific functional segments, respectively. ED is external force–induced directional factor. APP is a symmetric packing process.].

Figure 5.5 showed that nano-assembled systems are obtained through nano-assembly with three basic nano-assembly building units. Nano-assembled systems can have a variant range of structures and physical/chemical properties and diverse functional properties. For many nano-assembled systems, these general properties are those that are already known to other existing systems such as macroscopic counterparts. They can be straightly characterized. However, for many others, they can be novel properties that cannot be easily recognized and characterized. There are also nano-assembled systems whose general properties are overlapped by others. The concept of force balance for nano-assembly makes it possible for us to evaluate the specific properties that can be expected from certain nano-assembly building units. It can provide a nice in sight when choosing a proper nano-assembly route for a specific nano-assembled system and help clarify intended nanoscale properties (or nano-properties) with a reasonable degree of accuracy. Four elemental properties (which will be called nanoelements here after) for nano-assembled systems are proposed here in order to address these properties in a systematic manner.

They are nano-structural element, nano-property element, nano-mechanical element, and nano-communication element [45].

The symbols for each nanoelement are also shown in Fig. 5.6. Table 5.2 shows representative examples of the four nanoelements. A nano-structural element is the structural features that are inherited or designed from nano-assembly itself.

TABLE 5.2 Representative Examples of the Four Nanoelements N-SE, N-PE, N-ME, and N-CE Refer to Nano-structural, Nano-property, Nano-mechanical, and Nano-communication Elements, Respectively

	Nanoparticle		Gating and switching
N-SE	Nano-pore	N-ME	Rotation and oscillation
	Nano-film		Tweeze ring and fingering
	Nano-tube		Rolling and bearing
	Nano-rod		Self-directional movement
	Nano-hollow sphere		Capture and release
	Nanofabricated surface		Sensing
N-PE	Surface Plasmon	N-CE	Any macroscale performance by nano-integrated system and energy exchange which are performed by nano-machines
	Quantum size effect		
	Single electron tunneling		
	Surface catalytic activity		
	Mechanical strength		
	Energy conversion		
	Nano-confinement effect		

As shown in the Table 5.2, most of the nanostructure-based nano-assembled systems belong to this. A nano-property element is the properties that are inherited, induced, or designed from nano-assembly and its framework. Some of them could be the same properties as macroscale counterparts but in the nanoscale while others are those that emerge only when the systems have nanoscale features. A nano-mechanical element is the unit operations that are designed to express the motional aspects of nano-assembled systems. Finally, a nano-communication element is a signal, energy, or work that is designed to communicate with the macro world.

This nanoelement is almost exclusively for nanofabricated systems, nano-integrated systems, nano-devices, and nano-machines [46].

5.6.4 GENERAL ASSEMBLY DIAGRAM

The outcome of self-assembly is self-assembled aggregate. For nano-assembly, it goes one more step. The apparent initial outcome of nano-assembly is a nano-assembled system. But it is the nanoelements that make nano-assembled systems distinctive from self-assembled aggregates. Self-assembled aggregates have their own characteristic properties, which in many ways are effective, and many applications have been established using them over a wide range of scientific and technological fields. For nano-assembled systems, it is the nanoelements that define their characteristic properties, and with which we are seeking practical applications for nanotechnology systems [47].

Figure 5.7 presents the general rules of nano-assembly and their relationship with nanoelements. As a nano-assembly becomes more desired (moving toward the right hand direction on the horizontal arrow of attractive interaction–repulsive interaction balance), the nanoelement that will be expressed is a nano-structural element. Typical nano-pores, nanoparticles, nano-crystals, nano-emulsions, and nano-composites are more likely to be obtained on this side of the arrow. On the other hand, if a nano-assembly moves toward the left-hand side, it is more likely to obtain nano-assembled systems that usually need an aid of external force for their assembly. Colloidal crystal is one good example, especially when the size of nano-assembly building unit (colloidal particle) is increased. Many top-down operation-based nanoelements are other examples.

When a nano-assembly is involved with a directional interaction, the most likely nanoelement will be a film or surface-based nanoscale operation. Examples include most of the nano-structured films regardless of their detailed morphology. Nano-porous film, nano-layered film, and nano-patterned film are among them. It also includes most of the nanoscale products that are obtained as a result of directional growth (from the spherical-shape) such as nano-rods, nano-needles, and nanotubes. A good deal of nanofabrication is basically the nanoscale process that is performed on the surface, and thus becomes one prominent example for the upward direction on the vertical arrow. The opposite direction produces nanoelements, too. Some nanoparticles and nano-crystals can be obtained at this end. Most of the nano-property elements come along with nano-structural elements. And they are coupled to each other in many ways [48, 49].

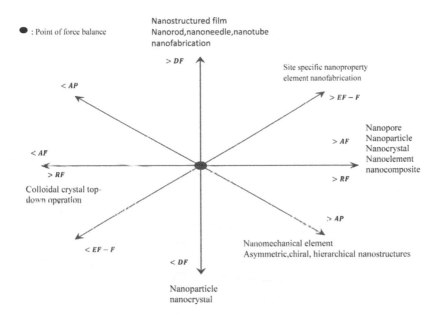

FIGURE 5.7 General rules of nano-assembly and different types of nanoelements AF, RF, and DF refer to attractive, repulsive, and directional forces, respectively. AP and EF-F are short for a symmetric packing and external force–specific functional segments, respectively.

Most of the nano-property elements originate because the nano-structural elements are in the nanoscale. And the changes in nano-property elements can be feasible because the changes in nano-structural elements are practical through nano-assembly. The nano-assemblies that occur with external force, specific functional and asymmetric packing segments are critical for nano-mechanical and nano-communication elements. Electron tunneling and Coulomb blockade are good examples. Nano-fabrication can take advantage of the unique features of external stimulus–specific nano-assembly, too. For a chiral nano-assembly, the chirality that is specific on each system can be used for the development of the nanostructures that can take advantage of the uniqueness, which includes highly asymmetric nanostructures, chiral nanoparticles, and some hierarchically constructed multiple-length-scale nanomaterials [50, 51].

It is also important for many unique types of nano-mechanical elements. By coupling with the external stimulus-specific nano-assembly, the development of nanoelements on this side (right-hand side of both external stimulus-specific and chiral nano-assemblies) can be much more fruitful. As far as the application for nanotechnology systems goes, the other side (left-hand side) of both diagonal arrows does not have much use in the development of specific nanoelements [52].

5.6.5   GENERAL TRENDS

Each nanofabricated system is a unique product of each fabrication system. Each nanoelement of the nano-fabricated system is a unique expression of its building units. They can be coupled locally or as a whole. They also can have a synergistic or an antagonistic outcome after the fabrication. All of these aspects have some degree of impact on the nanoelements of the nanofabricated systems. For some cases, different nanofabrication processes become the major reason for differentiating the nanoelements, even though the nanofabricated system might be the same [53].

Figure 5.8 shows a general trend of nanofabrication that covers these aspects from the three approaches. The mass assembling capability of nanofabrication becomes critically important when it goes to industrial scale. Generally, this capability is enhanced where fabrication is performed based on the bottom-up or the bottom-up/top-down hybrid approach. Because of the technical difficulties of top-down techniques and the limitations of the starting bulk materials that are comparable to them, the diversity of building units is much increased when the bottom-up or the hybrid approach is used. More diverse building units mean more diverse nanofabricated systems and more diverse nanoelements that can be explored. Structural diversity, hierarchy, and chirality are also important for widening the practicality of the nanofabricated systems [54].

It is easier to take advantage of two methods with a bottom-up or hybrid approach. Exact control of the building units is useful when the nanoelement is determined by the local control of a few main building units, as the top-down approach provides an advantage for this. Generally, the top-down approach has higher precision because of capability of manipulative assembly. Process ability means ease of the fabrication. This factor is important because it determines how practical a specific fabrication can be. It is, however, very much dependent on each nanofabrication system. Another factor is the structural integrity of the nanofabricated systems. It might appear that top-down processed systems would have better structural integrity because they are the products of the bulk materials. But structural integrity is measured not by the absolute strength of the nanofabricated systems but by their relative stability during actual use. As long as they perform the desired functions at given conditions, arguments about their absolute strength are less meaningful. Pure bottom-up fabrication in many cases provides enough; sometimes surprisingly strong, structural strength and resilience for nanofabricated systems to make them function properly even under harsh conditions [55, 56].

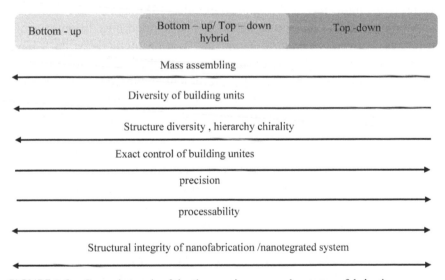

FIGURE 5.8 Generals trends of the three mains approaches to nanofabrication.

5.6.6 INVESTIGATION ON NEW STUDIES

The fields of nanoscience and nanotechnology generally concern the synthesis, fabrication and use of nanoelements and nanostructures at atomic, molecular and supra molecular levels. The nano-size of these elements and structures offers significant potential for research and applications across the scientific disciplines, including materials science, physics, chemistry, computer science, engineering and biology. Biological processes and methods, for example, are expected to be developed based entirely on nanoelements and their assembly into nanostructures. Other applications include developing nano-devices for use in semiconductors, electronics, photonics, optics, materials and medicine [57].

One class of nanoelements that has garnered considerable interest consists of carbon nanotubes. A carbon nanotube has a diameter on the order of nanometers and can be several micrometers in length. These nanoelements feature concentrically arranged carbon hexagons. Carbon nanotubes can behave as metals or semiconductors depending on their chirality and physical geometry. Other classes of nanoelements include, for example, nano-crystals, dendrimers, nanoparticles, nano-wires, biological materials, proteins, molecules and organic nanotubes [58].

Although carbon nanotubes have been assembled into different nanostructures, convenient nano-tools and fabrication methods to do it have not yet been developed. One obstacle has been the manipulation of individual nanoelements, which is often inefficient and tedious. This problem is particularly challenging when assembling complex nanostructures that require selecting and ordering millions of nanoelements across a large area [59].

To date, nanostructure assembly has focused on dispersing and manipulating nanoelements using atomic force or scanning tunneling microscopic methods. Although these methods are useful for fabricating simple nano-devices, neither is practical when selecting and patterning, for example, millions of nanoelements for more complex structures. The development of nano-machines or "nano-assemblers" which are programmed and used to order nanoelements for their assembly holds promise, although there have been few practical advancements with these machines.

The advancement of nanotechnology requires millions of nanoelements to be conveniently selected and simultaneously assembled. Three-dimensional nanostructure assembly also requires that nanoelements be ordered across a large area [60, 61].

Nanoelements have generated much interest due to their potential use in devices requiring nanoscale features such as new electronic devices, sensors, photonic crystals, advanced batteries, and many other applications. The realization of commercial applications, however, depends on developing high-rate and precise assembly techniques to place these elements onto desired locations and surfaces [62].

Different approaches have been used to carry out directed assembly of nanoelements in a desired pattern on a substrate, each approach having different advantages and disadvantages. In electrophoresis assembly, charged nanoelements are driven by an electric field onto a patterned conductor. This method is fast, with assembly typically taking less than a minute; however, it is limited to assembly on a conductive substrate [63]. Directed assembly can also be carried out onto a chemically functionalized surface. However, such assembly is a slow process, requiring up to several hours, because it is diffusion limited. Thus, there remains a need for a method of nanoelement assembly that is both rapid and not reliant on having either a conductive surface or a chemically functionalized surface [64].

In some cases, devices have a volume element having a larger diameter than the nanoelement arranged in epitaxial connection to the nanoelement. The volume element is being doped in order to provide a high charge carrier injection into the nanoelement and a low access resistance in an electrical connection. The nanoelement may be upstanding from a semiconductor substrate. A concentric layer of low resistivity material forms on the volume element forms a contact [65].

Semiconductor nanoelement devices show great promise, potentially outperforming standard electrical, opto-electrical, and sensor etc. semiconductor devices. These devices can use certain nanoelement specific properties, 2-D, 1-D, or 0-D quantum confinement, flexibility in axial material variation due to less lattice match restrictions, antenna properties, ballistic transport, wave guiding properties, etc. Furthermore, in order to design first-rate semiconductor devices from nanoelements, transistors, light emitting diodes, semiconductor lasers, and sensors, and to fabricate efficient contacts, particularly with low access resistance, to such devices, the ability to dope and fabricate doped regions is crucial [66, 67].

As an example the limitations in the commonly used planar technology are related to difficulties in making field effect transistors (FET), with low access resistance, the difficulty to control the threshold voltage in the post-growth process, the presence of short-channel effects as the planar gate length is reduced, and the lack of suitable substrate and lattice-matched hetero structure material for the narrow band gap technologies [68].

One advantage of a nanoelement FET is the possibility to tailor the band structure along the transport channel using segments of different band gap and or doping levels. This allows for a reduction in both the source-to-gate and gate-to-drain access resistance. These segments may be incorporated directly during the growth, which is not possible in the planar technologies. The doping of nanoelements is challenged by several factors. Physical incorporation of do pants into the nanoelement crystal may be inhibited, but also the established carrier concentration from a certain do pant concentration may be lowered as compared to the corresponding doped bulk semiconductor. One factor that limits the physical incorporation and solubility of do pants in nanoelements is that the nanoelement growth temperatures very often are moderate [69].

For vapor-liquid-solid (VLS) grown nanoelements, the solubility and diffusion of do pant in the catalytic particle will influence the do pant incorporation. One related effect, with similar long-term consequences, is the out-diffusion of do pants in the nanoelement to surface sites. Though not limited to VLS grown nanoelements, it is enhanced by the high surface to volume ratio of the nanoelement. Also, the efficiency of the doping, the amount of majority charge carriers established by ionization of donors/acceptor atoms at a certain temperature may be lowered compared to the bulk semiconductor, caused by an increase in donor or acceptor effective activation energy, due to the small dimensions of the nanoelement. Surface depletion effects, decreasing the volume of the carrier reservoir, will also be increased due to the high surface to volume ratio of the nanoelement [70].

The above described effects are not intended to establish a complete list, and the magnitudes of these effects vary with nanoelement material, do pant, and nanoelement dimensions. They may all be strong enough to severely decrease device performance.

KEYWORDS

- Ionization
- Mechanic Quantum
- Nanoelement
- Nanoscience
- Semiconductor nanoelement devices
- Wave function

CHAPTER 6

COMPUTATIONAL METHODS AND EVALUATION

CONTENTS

6.1   INTRODUCTION

Carbon nanotubes were first observed by Iijima, almost two decades ago [1], andsince then, extensive work has been carried out to characterize their properties [2–4]. A wide range of characteristic parameters has been reported for carbon nanotube nano-composites. There are contradictory reports that show the influence of carbon nanotubes on a particular property (e.g., Young's modulus) to be improving, indifferent or even deteriorating [5]. However, from the experimental point of view, it is a great challenge to characterize the structure and to manipulate the fabrication of polymer nano-composites. The development of such materials is still largely empirical and a finer degree of control of their properties cannot be achieved so far. Therefore, computer modeling and simulation will play an ever increasing role in predicting and designing material properties, and guiding such experimental work as synthesis and characterization, For polymer nano-composites, computer modeling and simulation are especially useful in the hierarchical characteristics of the structure and dynamics of polymer nano-composites ranging from molecular scale, micro scale to mesoscale and macroscale, in particular, the molecular structures and dynamics at the interface between nanoparticles and polymer matrix. The purpose of this review is to discuss the application of modeling and simulation techniques to polymer nano-composites. This includes a broad subject covering methodologies at various length and time scales and many aspects of polymer nano-composites. We organize the review as follows. In section 1 we will discuss about carbon nanotubes (CNTs) and nano-composite properties. In Section 6.2, we introduce briefly the computational methods used so far for the systems of polymer nano-composites which can be roughly divided into three types: molecular scale methods (e.g., molecular dynamics (MD), Monte Carlo (MC)), micro scale methods [e.g., Brownian dynamics (BD), dissipative particle dynamics (DPD), lattice Boltzmann (LB), time dependent Ginzburg–Lanau method, dynamic density functional theory (DFT) method], and mesoscale and macroscale methods [e.g., micromechanics, equivalent-continuum and self-similar approaches, finite element method (FEM)] [6] many researchers used this method for determine the mechanical properties of nanocomposite that in Section 6.3 will be discussed. In section 4 modeling of interfacial load transfer between CNT and polymer in nanocomposite will be introduced and finally we conclude the review by emphasizing the current challenges and future research directions.

6.2 CARBON NANOTUBE (CNT) AND NANO-COMPOSITE PROPERTIES

6.2.1 INTRODUCTION TO CNTS

CNTs are one dimensional carbon materials with aspect ratio greater than 1000. They are cylinders composed of rolled-up graphite planes with diameters in nanometer scale [7–10]. The cylindrical nanotube usually has at least one end capped with a hemisphere of fullerene structure. Depending on the process for CNT fabrication, there are two types of CNTs [8–11]: single-walled CNTs (SWCNTs) and multi walled CNTs (MWCNTs). SWCNTs consist of a single graphene layer rolled up into a seamless cylinder whereas MWCNTs consist of two or more concentric cylindrical shells of graphene sheets coaxially arranged around a central hollow core with Vander Waals forces between adjacent layers. According to the rolling angle of the graphene sheet, CNTs have three chiralities: arm chair, zigzag and chiral one. The tube chirality is defined by the chiral vector, Ch = na1 + ma2 (Fig. 6.1), where the integers (n, m) are the number of steps along the unit vectors (a1 and a2) of the hexagonal lattice [9, 10]. Using this (n, m) naming scheme, the three types of orientation of the carbon atoms around the nanotube circumference are specified. If n = m, the nanotubes are called "armchair." If m = 0, the nanotubes are called "zigzag." Otherwise, they are called "chiral." The chirality of nanotubes has significant impact on their transport properties, particularly the electronic properties. For a given (n, m) nanotube, if (2n + m) is a multiple of 3, then the nanotube is metallic, otherwise the nanotube is a semi conductor. Each MWCNT contains a multilayer of graphene, and each layer can have different chirality, so the prediction of its physical properties is more complicated than that of SWCNT. Figure 6.1 shows the CNT with different chirality.

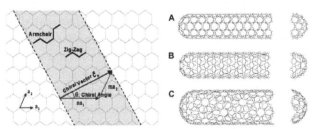

FIGURE 6.1 Schematic diagram showing how a hexagonal sheet of graphene is rolled to form a CNT with different chirality (A: armchair; B: zigzag; C: chiral).

6.2.2   CLASSIFICATION OF CNT/POLYMER NANO-COMPOSITES

Polymer composites, consisting of additives and polymer matrices, including thermoplastics, thermo sets and elastomers are considered to be an important group of relatively inexpensive materials for many engineering applications. Two or more materials are combined to produce composites that possess properties that are unique and cannot be obtained each material acting alone. For example, high modulus carbon fibers or silica particles are added into a polymer to produce reinforced polymer composites that exhibit significantly enhanced mechanical properties including strength, modulus and fracture toughness. However, there are some bottlenecks in optimizing the properties of polymer composites by employing traditional micron-scale fillers. The conventional filler content in polymer composites is generally in the range of 10–70wt%, which in turn results in a composite with a high density and high material cost. In addition, the modulus and strength of composites are often traded for high fracture toughness [12]. Unlike traditional polymer composites containing micron-scale fillers, the incorporation of nanoscale CNTs into a polymer system results in very short distance between the fillers, thus the properties of composites can be largely modified even at an extremely low content of filler. For example, the electrical conductivity of CNT/epoxy nano-composites can be enhanced several orders of magnitude with less than 0.5wt% of CNTs [13]. As described previously, CNTs are among the strongest and stiffest fibers ever known. These excellent mechanical properties combined with other physical properties of CNTs exemplify huge potential applications of CNT/polymer nano-composites. Ongoing experimental works in this area have shown some exciting results, although the much-anticipated commercial success has yet to be realized in the years ahead. In addition, CNT/polymer nano-composites are one of the most studied systems because of the fact that polymer matrix can be easily fabricated without damaging CNTs based on conventional manufacturing techniques, a potential advantage of reduced cost for mass production of nano-composites in the future. Following the first report on the preparation of a CNT/polymer nanocomposite in 1994 [14], many research efforts have been made to understand their structure–property relationship and find useful applications in different fields, and these efforts have become more pronounced after the realization of CNT fabrication in industrial scale with lower costs in the beginning of the twenty-first century [15]. According to the specific application, CNT/polymer nano-composites can be classified as structural or functional composites [16]. For the structural composites, the unique mechanical properties of CNTs, such as the high modulus, tensile strength and strain to fracture, are explored to obtain structural materials with much improved mechanical properties. As for CNT/polymer functional composites many other unique properties of CNTs, such as electrical, thermal optical and damping properties along with their excellent mechanical properties, are used to develop multifunctional composites for applications in the fields of heat resistance, chemical sensing, electrical and thermal management, photoemission, electromagnetic absorbing and energy storage performances, etc.

6.3 MODELING AND SIMULATION TECHNIQUES

6.3.1 MOLECULAR SCALE METHODS

The modeling and simulation methods at molecular level usually employ atoms, molecules or their clusters as the basic units considered. The most popular methods include molecular mechanics (MM), MD and MC simulation. Modeling of polymer nano-composites at this scale is predominantly directed toward the thermodynamics and kinetics of the formation, molecular structure and interactions. The diagram in Fig. 6.1 describes the equation of motion for each method and the typical properties predicted from each of them [17–22]. Two widely used molecular scale methods, namely MD and MC, are introduced and discussed in the following subsections.

6.3.1.1 MOLECULAR DYNAMICS

MD is a computer simulation technique that allows one to predict the time evolution of a system of interacting particles (e.g., atoms, molecules, granules, etc.) and estimate the relevant physical properties [23, 24]. Specifically, it generates such information as atomic positions, velocities and forces from which the macroscopic properties (e.g., pressure, energy, heat capacities) can be derived by means of statistical mechanics. MD simulation usually consists of three constituents: (i) a set of initial conditions (e.g., initial positions and velocities of all particles in the system); (ii) the interaction potentials to represent the forces among all the particles; (iii) the evolution of the system in time by solving a set of classical Newtonian equations of motion for all particles in the system. The equation of motion is generally given by

$$\vec{F}_1(t) = m_i \frac{d^2 \vec{r}_1}{dt^2} \tag{1}$$

where $\vec{F}_i(t)$ is the force acting on the it H atom or particle at time t which is obtained as the negative gradient of the interaction potential U, m_i is the atomic mass and $\vec{r}_i$ the atomic position. A physical simulation involves the proper selection of interaction potentials, numerical integration, periodic boundary conditions, and the controls of pressure and temperature to mimic physically meaningful thermodynamic ensembles. The interaction potentials together with their parameters, that is, the so-called force field, describe in detail how the particles in a system interact with each other, that is, how the potential energy of a system depends on the particle coordinates. Such a force field may be obtained by quantum method (e.g., AB initio), empirical method (e.g., Lennard-Jones, Mores, and Born-Mayer) or quantum-empirical method (e.g., embedded atom model, glue model, bond order potential). The criteria for selecting a force field include the accuracy, transferability and computational speed. A typical interaction potential U may consist of a number of bonded and non-bonded interaction terms:

$$U(\vec{r}_1,\vec{r}_2,\vec{r}_3,...,\vec{r}_n) = \sum_{i_{bond}}^{N_{bond}} U_{bond}(i_{bond},\vec{r}_a,\vec{r}_b) + \sum_{i_{angle}}^{N_{angle}} U_{angle}(i_{angle},\vec{r}_a,\vec{r}_b,\vec{r}_c) + \sum_{i_{torsion}}^{N_{torsion}} U_{torsion}(i_{torsion},\vec{r}_a,\vec{r}_b,\vec{r}_c,\vec{r}_d)$$

$$+ \sum_{i_{inversion}}^{N_{inversion}} U_{inversion}(i_{inversion},\vec{r}_a,\vec{r}_b,\vec{r}_c,\vec{r}_d) + \sum_{i=1}^{N-1}\sum_{j>i}^{N} U_{vdw}(i,j,\vec{r}_a,\vec{r}_b) + \sum_{i=1}^{n-1}\sum_{j>i}^{N} U_{electrostatic}(i,j,\vec{r}_a,\vec{r}_b) \tag{2}$$

The first four terms represent bonded interactions, that is, bond stretching U bond, bond-angle bend U angle and dihedral angle torsion U torsion and inversion interaction U inversion, while the last two terms are non-bonded interactions, that is, Vander Waals energy U vdW and electrostatic energy U electrostatic. In the equation, $\vec{r}_a,\vec{r}_b,\vec{r}_c,\vec{r}_d$ are the positions of the atoms or particles specifically involved in a given interaction; $N_{bond}, N_{angle}, N_{torsion}$ and $N_{inversion}$ stand for the total numbers of these respective interactions in the simulated system; $i_{bond}, i_{angle}, i_{torsion}$ and $i_{inversion}$ uniquely specify an individual interaction of each type; i and j in the Vander Waals and electrostatic terms indicate the atoms involved in the interaction. There are many algorithms for integrating the equation of motion using finite difference methods. The algorithms of varlet, velocity varlet, leap-frog and Bee man, are commonly used in MD simulations [23]. All algorithms assume that the atomic position $\vec{r}$, velocities $\vec{v}$ and accelerations $\vec{a}$ can be approximated by a Taylor series expansion:

$$\vec{r}(t+\delta t) = \vec{r}(t) + \vec{v}(t)\delta t + \frac{1}{2}\vec{a}(t)\delta^2 t + ... \tag{3}$$

$$\vec{v}(t+\delta t) = \vec{v}(t)\delta t + \frac{1}{2}\vec{b}(t)\delta^2 t + ... \tag{4}$$

$$\vec{a}(t+\delta t) = \vec{a}(t) + \vec{b}(t)\delta t + ... \tag{5}$$

Generally speaking, a good integration algorithm should conserve the total energy and momentum and be time-reversible. It should also be easy to implement and computationally efficient, and permit a relatively long time step. The Verlet algorithm is probably the most widely used method. It uses the positions $\vec{r}(t)$ and accelerations $\vec{a}(t)$ at time t, and the positions $\vec{r}(t-\delta t)$ from the previous step (--δ) to calculate the new positions $\vec{r}(t+\delta t)$ at (t + δt), we have:

$$\vec{r}(t+\delta t) = \vec{r}(t) + \vec{v}(t)\delta t + \frac{1}{2}\vec{a}(t)\delta t^2 + ... \tag{6}$$

$$\vec{r}(t-\delta t) = \vec{r}(t) - \vec{v}(t)\delta t + \frac{1}{2}\vec{a}(t)\delta t^2 + ... \tag{7}$$

$$\vec{r}(t+\delta t) = 2\vec{r}(t)\delta t - \vec{r}(t-\delta t) + \vec{a}(t)\delta t^2 + ... \tag{8}$$

The velocities at time t and $t+\frac{1}{2\delta t}$ can be respectively estimated

$$\vec{v}(t) = \left[\vec{r}(t+\delta t) - \vec{r}(t-\delta t)\right]/2\delta t \tag{9}$$

$$\vec{v}(t+1/2\delta t) = \left[\vec{r}(t+\delta t) - \vec{r}(t-\delta t)\right]/\delta t \tag{10}$$

MD simulations can be performed in many different ensembles, such as grand canonical (μVT), micro canonical (NVE), canonical (NVT) and isothermal–isobaric (NPT). The constant temperature and pressure can be controlled by adding an appropriate thermostat (e.g., Berendsen, Nose, Nose–Hoover and Nose–Poincare) and barostat (e.g., Andersen, Hoover and Berendsen), respectively. Applying MD into polymer composites allows us to investigate into the effects of fillers on polymer structure and dynamics in the vicinity of polymer–filler interface and also to probe the effects of polymer–filler interactions on the materials properties.

6.3.1.2   MONTE CARLO

MC technique, also called Metropolis method [24], is a stochastic method that uses random numbers to generate a sample population of the system from which one can calculate the properties of interest. A MC simulation usually consists of three typical steps. In the first step, the physical problem under investigation is translated into an analogous probabilistic or statistical model. In the second step, the probabilistic model is solved by a numerical stochastic sampling experiment. In the third step, the obtained data are analyzed by using statistical methods. MC provides only the information on equilibrium properties (e.g., free energy, phase equilibrium), different from MD, which gives nonequilibrium, as well as equilibrium properties. In a NVT ensemble with N atoms, one hypothesizes a new configuration by arbitrarily or systematically moving one atom from position i→j. Due to such atomic movement, one can compute the change in the system Hamiltonian ΔH:

$$\Delta H = H(j) - H(i) \tag{11}$$

where H(i) and H(j) are the Hamiltonian associated with the original and new configuration, respectively.

This new configuration is then evaluated according to the following rules. If $\Delta H < 0$ then the atomic movements would bring the system to a state of lower energy. Hence, the movement is immediately accepted and the displaced atom remains in its new position. If $\Delta H \geq 0$, the move is accepted only with a certain probability $P_{i \to j}$ which is given by

$$Pi \rightarrow j \propto \exp(-\frac{\Delta H}{K_B T}) \qquad (12)$$

where K_B is the Boltzmann constant. According to Metropolis et al. [25] one can generate a random number ζ between 0 and 1 and determine the new configuration according to the following rule:

$$\xi \leq \exp(-\frac{\Delta H}{K_B T}); \text{ The move is accepted;} \qquad (13)$$

$$\xi \rangle \exp(-\frac{\Delta H}{K_B T}); \text{ The move is not accepted.} \qquad (14)$$

If the new configuration is rejected, one counts the original position as a new one and repeats the process by using other arbitrarily chosen atoms. In a μVT ensemble, one hypothesizes a new configuration j by arbitrarily choosing one atom and proposing that it can be exchanged by an atom of a different kind. This procedure affects the chemical composition of the system. Also, the move is accepted with a certain probability. However, one computes the energy change ΔU associated with the change in composition. The new configuration is examined according to the following rules. If $\Delta U < 0$, the moves of compositional changes is accepted. However, if $\Delta U \geq 0$, the move is accepted with a certain probability which is given by

$$Pi \rightarrow j \propto \exp(-\frac{\Delta U}{K_B T}) \qquad (15)$$

where ΔU is the change in the sum of the mixing energy and the chemical potential of the mixture. If the new configuration is rejected one counts the original configuration as a new one and repeats the process by using some other arbitrarily or systematically chosen atoms. In polymer nano-composites, MC methods have been used to investigate the molecular structure at nanoparticle surface and evaluate the effects of various factors.

6.3.2 MICRO SCALE METHODS

The modeling and simulation at micro scale aim to bridge molecular methods and continuum methods and avoid their shortcomings. Specifically, in nanoparticle–polymer systems, the study of structural evolution (i.e., dynamics of phase separation) involves the description of bulk flow (i.e., hydrodynamic behavior) and the interactions between nanoparticle and polymer components. Note that hydrodynamic behavior is relatively straightforward to handle by continuum methods but is very difficult and expensive to treat by atomistic methods. In contrast, the interactions between components can be examined at an atomistic level but are usually not straightforward to incorporate at the continuum level. Therefore, various simulation methods have been evaluated and extended to study the microscopic structure and phase separation of these polymer nanocomposites, including BD, DPD, LB, time-dependent Ginsburg–Landau (TDGL) theory, and dynamic DFT. In these methods, a polymer system is usually treated with a field description or microscopic particles that incorporate molecular details implicitly. Therefore, they are able to simulate the phenomena on length and time scales currently inaccessible by the classical MD methods.

6.3.2.1 BROWNIAN DYNAMICS

BD simulation is similar to MD simulations [26]. However; it introduces a few new approximations that allow one to perform simulations on the microsecond timescale whereas MD simulation is known up to a few nanoseconds. In BD the explicit description of solvent molecules used in MD is replaced with an implicit continuum solvent description. Besides, the internal motions of molecules are typically ignored, allowing a much larger time step than that of MD. Therefore, BD is particularly useful for systems where there is a large gap of time scale governing the motion of different components. For example, in polymer–solvent mixture, a short time-step is required to resolve the fast motion of the solvent molecules, whereas the evolution of the slower modes of the system requires a larger time step. However, if the detailed motion of the solvent molecules is concerned, they may be removed from the simulation and their effects on the polymer are represented by dissipative ($-\gamma$ P) and random (σ $\zeta(t)$) force terms. Thus, the forces in the governing Eq. (16) is replaced by a Langevin equation,

$$F_i(t) = \sum_{i \neq j} F_{ij}^{\ c} - \gamma P_i + \sigma \zeta_i(t) \tag{16}$$

where F_{ij}^c is the conservative force of particle j acting on particle i, γ and σ are constants depending on the system, P_i the momentum of particle i, and $\varsigma(t)$ a Gaussian random noise term. One consequence of this approximation of the fast degrees of freedom by fluctuating forces is that the energy and momentum are no longer conserved, which implies that the macroscopic behavior of the system will not be hydrodynamic. In addition, the effect of one solute molecule on another through the flow of solvent molecules is neglected. Thus, BD can only reproduce the diffusion properties but not the hydrodynamic flow properties since the simulation does not obey the Navier–Stokes equations.

6.3.2.2 DISSIPATIVE PARTICLE DYNAMICS

DPD was originally developed by Hoogerbrugge and Koelman [27]. It can simulate both Newtonian and non-Newtonian fluids, including polymer melts and blends, on microscopic length and time scales. Like MD and BD, DPD is a particle-based method. However, its basic unit is not a single atom or molecule but a molecular assembly (i.e., a particle). DPD particles are defined by their mass M_i, position r_i and momentum P_i. The interaction force between two DPD particles i and j can be described by a sum of conservative F_{ij}^c, dissipative F_{ij}^d and random forces F_{ij}^R [28–30]:

$$F_{ij} = F_{ij}^{\,C} + F_{ij}^{\,D} + F_{\,ij}^{R} \qquad (17)$$

While the interaction potentials in MD are high-order polynomials of the distance r_{ij} between two particles, in DPD the potentials are softened so as to approximate the effective potential at microscopic length scales. The form of the conservative force in particular is chosen to decrease linearly with increasing r_{ij}. Beyond a certain cut-off separation r_c, the weight functions and thus the forces are all zero. Because the forces are pair wise and momentum is conserved, the macroscopic behavior directly incorporates Navier–Stokes hydrodynamics. However, energy is not conserved because of the presence of the dissipative and random force terms, which are similar to those of BD, but incorporate the effects of Brownian motion on larger length scales. DPD has several advantages over MD, for example, the hydrodynamic behavior is observed with far fewer particles than required in a MD simulation because of its larger particle size. Besides, its force forms allow larger time steps to be taken than those in MD.

6.3.2.3   LATTICE BOLTZMANN

LB [31] is another micro scale method that is suited for the efficient treatment of polymer solution dynamics. It has recently been used to investigate the phase separation of binary fluids in the presence of solid particles. The LB method is originated from lattice gas automaton, which is constructed as a simplified, fictitious molecular dynamic in which space, time and particle velocities are all discrete. A typical lattice gas automaton consists of a regular lattice with particles residing on the nodes. The main feature of the LB method is to replace the particle occupation variables (Boolean variables), by single-particle distribution functions (real variables) and neglect individual particle motion and particle–particle correlations in the kinetic equation. There are several ways to obtain the LB equation from either the discrete velocity model or the Boltzmann kinetic equation, and to derive the macroscopic Navier–Stokes equations from the LB equation. An important advantage of the LB method is that microscopic physical interactions of the fluid particles can be conveniently incorporated into the numerical model. Compared with the Navier–Stokes equations, the LB method can handle the interactions among fluid particles and reproduce the micro scale mechanism of hydrodynamic behavior. Therefore, it belongs to the MD in nature and bridges the gap between the molecular level and macroscopic level. However, its main disadvantage is that it is typically not guaranteed to be numerically stable and may lead to physically unreasonable results, for instance, in the case of high forcing rate or high inter particle interaction strength.

6.3.2.4   TIME-DEPENDENT GINZBURG-LANDAU METHOD

TDGL is a micro scale method for simulating the structural evolution of phase-separation in polymer blends and block copolymers. It is based on the Cahn–Hilliard–Cook (CHC) nonlinear diffusion equation for a binary blend and falls under the more general phase-field and reaction-diffusion models [32–34]. In the TDGL method, a free-energy function is minimized to simulate a temperature quench from the miscible region of the phase diagram to the immiscible region. Thus, the resulting time-dependent structural evolution of the polymer blend can be investigated by solving the TDGL/CHC equation for the time dependence of the local blend concentration. Scientists have discussed and applied this method to polymer blends and particle-filled polymer systems [35]. This model reproduces the growth kinetics of the TDGL model, demonstrating that such quantities are insensitive to the precise form of the double-well potential of the bulk free-energy term. The TDGL and CDM methods have recently been used to investigate the phase-separation of polymer nano-composites and polymer blends in the presence of nanoparticles [36–40].

6.3.2.5 DYNAMIC DFT METHOD

Dynamic DFT method is usually used to model the dynamic behavior of polymer systems and has been implemented in the software package Mesodyn™ from Accelrys [41]. The DFT models the behavior of polymer fluids by combining Gaussian mean-field statistics with a TDGL model for the time evolution of conserved order parameters. However, in contrast to traditional phenomenological free-energy expansion methods employed in the TDGL approach, the free energy is not truncated at a certain level, and instead retains the full polymer path integral numerically. At the expense of a more challenging computation, this allows detailed information about a specific polymer system beyond simply the Flory–Huggins parameter and mobility to be included in the simulation. In addition, viscoelasticity, which is not included in TDGL approaches, is included at the level of the Gaussian chains. A similar DFT approach has been developed by Doi et al. [42, 43] and forms the basis for their new software tool Simulation Utilities for Soft and Hard Interfaces (SUSHI), one of a suite of molecular and mesoscale modeling tools (called OCTA) developed for the simulation of polymer materials [44]. The essence of dynamic DFT method is that the instantaneous unique conformation distribution can be obtained from the off-equilibrium density profile by coupling a fictitious external potential to the Hamiltonian. Once such distribution is known, the free energy is then calculated by standard statistical thermodynamics. The driving force for diffusion is obtained from the spatial gradient of the first functional derivative of the free energy with respect to the density. Here, we describe briefly the equations for both polymer and particle in the di block polymer-particle composites [38].

6.3.3   MESOSCALE AND MACROSCALE METHODS

Despite the importance of understanding the molecular structure and nature of materials, their behavior can be homogenized with respect to different aspects which can be at different scales. Typically, the observed macroscopic behavior is usually explained by ignoring the discrete atomic and molecular structure and assuming that the material is continuously distributed throughout its volume. The continuum material is thus assumed to have an average density and can be subjected to body forces such as gravity and surface forces. Generally speaking, the macroscale methods (or called continuum methods hereafter) obey the fundamental laws of: (i) continuity, derived from the conservation of mass; (ii) equilibrium, derived from momentum considerations and Newton's second law; (iii) the moment of momentum principle, based on the model that the time rate of change of angular momentum with respect to an arbitrary point is equal to the resultant moment; (iv) conservation of energy, based on the first law of thermodynamics; and (v) conservation of entropy, based on the second law of thermodynamics. These laws provide the basis for the continuum model and must be coupled with the appropriate constitutive equations and the equations of state to provide all the equations necessary for solving a continuum problem. The continuum method relates the deformation of a continuous medium to the external forces acting on the medium and the resulting internal stress and strain. Computational approaches range from simple closed-form analytical expressions to micromechanics and complex structural mechanics calculations based on beam and shell theory. In this section, we introduce some continuum methods that have been used in polymer nano-composites, including micromechanics models (e.g., Halpin–Tsai model, Mori–Tanaka model), equivalent-continuum model, self-consistent model and finite element analysis.

6.3.4   MICROMECHANICS

Since the assumption of uniformity in continuum mechanics may not hold at the micro scale level, micromechanics methods are used to express the continuum quantities associated with an infinitesimal material element in terms of structure and properties of the micro constituents. Thus, a central theme of micromechanics models is the development of a representative volume element (RVE) to statistically represent the local continuum properties. The RVE is constructed to ensure that the length scale is consistent with the smallest constituent that has a first-order effect on the macroscopic behavior. The RVE is then used in a repeating or periodic nature in the full-scale model. The micromechanics method can account for interfaces between constituents, discontinuities, and coupled mechanical and nonmechanical properties. Our purpose is to review the micromechanics methods used for polymer nano-composites. Thus, we only discuss here some important concepts of micromechanics as well as the Halpin-Tsai model and Mori-Tanaka model.

6.3.4.1   BASIC CONCEPTS

When applied to particle reinforced polymer composites, micromechanics models usually follow such basic assumptions as (i) linear elasticity of fillers and polymer matrix; (ii) the fillers are ax symmetric, identical in shape and size, and can be characterized by parameters such as aspect ratio; (iii) well-bonded filler–polymer interface and the ignorance of interfacial slip, filler–polymer debonding or matrix cracking. The first concept is the linear elasticity, that is, the linear relationship between the total stress and infinitesimal strain tensors for the filler and matrix as expressed by the following constitutive equations:

$$\text{For filler } \sigma^f = C^f \varepsilon^f \tag{18}$$

$$\text{For matrix } \sigma^m = C^m \varepsilon^m \tag{19}$$

where C is the stiffness tensor. The second concept is the average stress and strain. Since the point wise stress field $\sigma(x)$ and the corresponding strain field $\varepsilon(x)$ are usually nonuniform in polymer composites, the volume–average stress $\bar{\sigma}$ and strain $\bar{\sigma}$ are then defined over the representative averaging volume V, respectively.

$$\bar{\sigma} = \frac{1}{V} \int \sigma(x)dv \tag{20}$$

$$\bar{\varepsilon} = \frac{1}{V} \int \varepsilon(x)dv \tag{21}$$

Therefore, the average filler and matrix stresses are the averages over the corresponding volumes v_f and v_m, respectively.

$$\bar{\sigma}_f = \frac{1}{V_f} \int \sigma(x)dv \tag{22}$$

$$\bar{\sigma}_m = \frac{1}{V_m} \int \sigma(x)dv \tag{23}$$

The average strains for the fillers and matrix are defined, respectively, as

$$\bar{\varepsilon}_f = \frac{1}{V_f} \int \varepsilon(x)dv \tag{24}$$

$$\bar{\varepsilon}_m = \frac{1}{V_m} \int \varepsilon(x)dv \tag{25}$$

Based on the above definitions, the relationships between the filler and matrix averages and the overall averages can be derived as follows:

$$\bar{\sigma} = \bar{\sigma}_f v_f + \bar{\sigma}_m v_m \tag{26}$$

$$\bar{\varepsilon} = \bar{\varepsilon}_f v_f + \bar{\varepsilon}_m v_m \tag{27}$$

where v_f, v_m are the volume fractions of the fillers and matrix, respectively.

The third concept is the average properties of composites, which are actually the main goal of a micromechanics model. The average stiffness of the composite is the tensor C that maps the uniform strain to the average stress

$$\bar{\sigma} = \bar{\varepsilon}C \tag{28}$$

The average compliance S is defined in the same way:

$$\bar{\varepsilon} = \bar{\sigma}S \tag{29}$$

Another important concept is the strain–concentration and stress–concentration tensors A and B, which are basically the ratios between the average filler strain (or stress) and the corresponding average of the composites.

$$\overline{\varepsilon_f} = \bar{\varepsilon}A \tag{30}$$

$$\overline{\sigma_f} = \bar{\sigma}B \tag{31}$$

Using the above concepts and equations, the average composite stiffness can be obtained from the strain concentration tensor A and the filler and matrix properties:

$$C = C_m + v_f(C_f - C_m)A \tag{32}$$

6.3.4.2   HALPIN–TSAI MODEL

The Halpin–Tsai model is a well-known composite theory to predict the stiffness of unidirectional composites as a functional of aspect ratio. In this model, the longitudinal E11 and transverse E22 engineering moduli are expressed in the following general form:

$$\frac{E}{E_m} = \frac{1 + \zeta \eta v_f}{1 - \eta v_f} \tag{33}$$

where E and E_m represent the Young's modulus of the composite and matrix, respectively, v_f is the volume fraction of filler, and η is given by:

$$\eta = \frac{\dfrac{E}{E_m} - 1}{\dfrac{E_f}{E_m} + \zeta_f} \tag{34}$$

where E_f represents the Young's modulus of the filler and ζ_f the shape parameter depending on the filler geometry and loading direction. When calculating longitudinal modulus E_{11}, ζ_f is equal to l/t, and when calculating transverse modulus E_{22}, ζ_f is equal to w/t. Here, the parameters of l, w and t are the length, width and thickness of the dispersed fillers, respectively. If $\zeta_f \to 0$, the Halpin–Tsai theory converges to the inverse rule of mixture (lower bound):

$$\frac{1}{E} = \frac{v_f}{E_f} + \frac{1 - v_f}{E_m} \tag{35}$$

Conversely, if $\zeta_f \to \infty$, the theory reduces to the rule of mixtures (upper bound),

$$E = E_f v_f + E_m (1 - v_f) \tag{36}$$

6.3.4.3 MORI– TANAKA MODEL

The Mori–Tanaka model is derived based on the principles of Eshelby's inclusion model for predicting an elastic stress field in and around ellipsoidal filler in an infinite matrix. The complete analytical solutions for longitudinal E_{11} and transverse E_{22} elastic moduli of an isotropic matrix filled with aligned spherical inclusion are [45, 46]:

$$\frac{E_{11}}{E_m} = \frac{A_0}{A_0 + v_f (A_1 + 2v_0 A_2)} \tag{37}$$

$$\frac{E_{22}}{E_m} = \frac{2A_0}{2A_0 + v_f (-2A_3 + (1 - v_0 A_4) + (1 + v_0) A_5 A_0)} \tag{38}$$

where E_m represents the Young's modulus of the matrix, v_f the volume fraction of filler, v_0 the Poisson's ratio of the matrix, parameters, A0, A1, …, A5 are functions of the Eshelby's tensor and the properties of the filler and the matrix, including Young's modulus, Poisson's ratio, filler concentration and filler aspect ratio [45].

6.3.4.5 EQUIVALENT-CONTINUUM AND SELF-SIMILAR APPROACHES

Numerous micromechanical models have been successfully used to predict the macroscopic behavior of fiber-reinforced composites. However, the direct use of these models for nanotube-reinforced composites is doubtful due to the significant scale difference between nanotube and typical carbon fiber. Recently, two methods have been proposed for modeling the mechanical behavior of single walled carbon nanotube (SWCN) composites: equivalent-continuum approach and self-similar approach [47]. The equivalent-continuum approach was proposed by Odegard et al. [48]. In this approach, MD was used to model the molecular interactions between SWCN–polymer and a homogeneous equivalent-continuum reinforcing element (e.g., a SWCN surrounded bpolymer) was constructed as shown in Fig. 6.2. Then, micromechanics are used to determine the effective bulk properties of the equivalent-continuum reinforcing element embedded in a continuous polymer. The equivalent-continuum approach consists of four major steps, as briefly described below.

Step 1: MD simulation is used to generate the equilibrium structure of a SWCN–polymer composite and then to establish the RVE of the molecular model and the equivalent-continuum model.

Step 2: The potential energies of deformation for the molecular model and effective fiber are derived and equated for identical loading conditions. The bonded and nonbonded interactions within a polymer molecule are quantitatively described by MM. For the SWCN/polymer system, the total potential energy U^m of the molecular model is:

$$U^m = \sum U^r(K_r) + \sum U^\theta(K_\theta) + \sum U^{vdw}(K_{vdw}) \qquad (39)$$

where U^r, U^θ and U^{vdw} are the energies associated with covalent bond stretching, bond-angle bending, and Vander Waals interactions, respectively. An equivalent-truss model of the RVE is used as an intermediate step to link the molecular and equivalent-continuum models. Each atom in the molecular model is represented by a pin-joint, and each truss element represents an atomic bonded or nonbonded interaction. The potential energy of the truss model is

$$U^t = \sum U^a(E^a) + \sum U^b(E^b) + \sum U^c(E^c) \tag{40}$$

where U^a, U^b and U^c are the energies associated with truss elements that represent covalent bond stretching, bond-angle bending, and Vander Waals interactions, respectively. The energies of each truss element are a function of the Young's modulus, E.

Step 3: A constitutive equation for the effective fiber is established. Since the values of the elastic stiffness tensor components are not known a priori, a set of loading conditions are chosen such that each component is uniquely determined from

$$U^f = U^t = U^m \tag{41}$$

Step 4: Overall constitutive properties of the dilute and unidirectional SWCN/polymer composite are determined with Mori–Tanaka model with the mechanical properties of the effective fiber and the bulk polymer. The layer of polymer molecules that are near the polymer/nano-tube interface (Fig. 6.2) is included in the effective fiber and it is assumed that the matrix polymer surrounding the effective fiber has mechanical properties equal to those of the bulk polymer. The self-similar approach was proposed by Pipes and Hubert [49], which consists of three major steps:

First, a helical array of SWCNs is assembled. This array is termed as the SWCN nano-array where 91 SWCNs make up the cross-section of the helical nano-array. Then, the SWCN nano-arrays is surrounded by a polymer matrix and assembled into a second twisted array, termed as the SWCN nano-wire Finally, the SWCN nano-wires are further impregnated with a polymer matrix and assembled into the final helical array—the SWCN microfiber. The self-similar geometries described in the nano-array, nano-wire and microfiber (Fig. 6.3) allow the use of the same mathematical and geometric model for all three geometries [49].

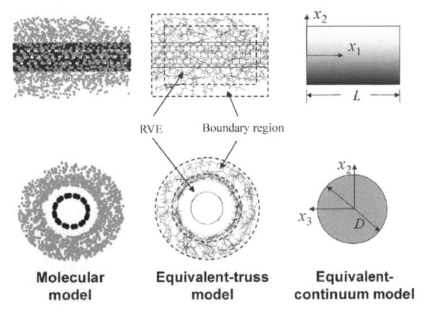

Molecular model   **Equivalent-truss model**   **Equivalent-continuum model**

FIGURE 6.2 Equivalent-continuum modeling of effective fiber [48].

6.3.4.6 FINITE ELEMENT METHOD

FEM is a general numerical method for obtaining approximate solutions in space to initial-value and boundary-value problems including time-dependent processes. It employs preprocessed mesh generation, which enables the model to fully capture the spatial discontinuities of highly inhomogeneous materials. It also allows complex, nonlinear tensile relationships to be incorporated into the analysis. Thus, it has been widely used in mechanical, biological and geological systems. In FEM, the entire domain of interest is spatially discredited into an assembly of simply shaped sub domains (e.g., hexahedra or tetrahedral in three dimensions, and rectangles or triangles in two dimensions) without gaps and without overlaps. The sub domains are interconnected at joints (i.e., nodes). The implementation of FEM includes the important steps shown in Fig. 6.4. The energy in FEM is taken from the theory of linear elasticity and thus the input parameters are simply the elastic moduli and the density of the material. Since these parameters are in agreement with the values computed by MD, the simulation is consistent across the scales. More specifically, the total elastic energy in the absence of tractions and body forces within the continuum model is given by [50]:

$$U = U_v + U_k \tag{42}$$

$$U_k = 1/2 \int dr p(r) \left| \dot{U}_r \right|^2 \tag{43}$$

$$U_v = \frac{1}{2} \int dr \sum\nolimits_{\mu,\nu,\lambda,\sigma=1}^{3} \varepsilon_{\mu\nu}(r) C_{\mu\nu\lambda\sigma} \lambda\sigma(r) \tag{44}$$

where U_v is the Hookian potential energy term which is quadratic in the symmetric strain tensor e, contracted with the elastic constant tensor C. The Greek indices (i.e., m, n, l, s) denote Cartesian directions. The kinetic energy U_k involves the time rate of change of the displacement field $\dot{U}$, and the mass density ρ.

These are fields defined throughout space in the continuum theory. Thus, the total energy of the system is an integral of these quantities over the volume of the sample dυ. The FEM has been incorporated in some commercial software packages and open source codes (e.g., ABAQUS, ANSYS, Palmyra and OOF) and widely used to evaluate the mechanical properties of polymer composites. Some attempts have recently been made to apply the FEM to nanoparticle-reinforced polymer nano-composites. In order to capture the multi scale material behaviors, efforts are also underway to combine the multi scale models spanning from molecular to macroscopic levels [51, 52].

6.4 MULTI SCALE MODELING OF MECHANICAL PROPERTIES

In Odegard's study [48], a method has been presented for linking atomistic simulations of nano-structured materials to continuum models of the corresponding bulk material. For a polymer composite system reinforced with single walled carbon nanotubes (SWNT), the method provides the steps whereby the nanotube, the local polymer near the nanotube, and the nanotube/polymer interface can be modeled as an effective continuum fiber by using an equivalent-continuum model. The effective fiber retains the local molecular structure and bonding information, as defined by molecular dynamics, and serves as a means for linking the equivalent-continuum and micromechanics models. The micromechanics method is then available for the prediction of bulk mechanical properties of SWNT/polymer composites as a function of nanotube size, orientation, and volume fraction. The utility of this method was examined by modeling tow composites that both having an interface. The elastic stiffness constants of the composites were determined for both aligned and three-dimensional randomly oriented nanotubes, as a function of nanotube length and volume fraction. They used Mori–Tanaka model [53] for random and oriented fibers position and compare their model with mechanical properties, the interface between fiber and matrix was assumed perfect. Motivated by micrographs showing that embedded nanotubes often exhibit significant curvature within the polymer, Fisher et al. [54] have developed a model combining finite element results and micromechanical methods (Mori-Tanaka) to determine the effective reinforcing modulus of a wavy embedded nanotube with perfect bonding and random fiber orientation assumption. This effective reinforcing modulus (ERM) is then used within a multiphase micromechanics model to predict the effective modulus of a polymer reinforced with a distribution of wavy nanotubes. We found that even slight nanotube curvature significantly reduces the effective reinforcement when compared to straight nanotubes. These results suggest that nanotube waviness may be an additional mechanism limiting the modulus enhancement of nanotube-reinforced polymers. Bradshaw et al. [55] investigated the degree to which the characteristic waviness of nanotubes embedded in polymers can impact the effective stiffness of these materials. A 3D finite element model of a single infinitely long sinusoidal fiber within an infinite matrix is used to numerically compute the dilute strain concentration tensor. A Mori–Tanaka model uses this tensor to predict the effective modulus of the material with aligned or randomly oriented inclusions. This hybrid finite element micromechanical modeling technique is a powerful extension of general micromechanics modeling and can be applied to any composite microstructure containing nonellipsoidal inclusions. The results demonstrate that nanotube waviness results in a reduction of the effective modulus of the composite relative to straight nanotube reinforcement. The degree of reduction is dependent on the ratio of the sinusoidal wavelength to the nanotube diameter. As this wavelength ratio increases, the effective stiffness of a composite with randomly oriented wavy nanotubes converges to the result obtained with straight nanotube inclusions.

The effective mechanical properties of carbon nanotube-based composites are evaluated by Liu and Chen [56] using a 3-D nanoscale RVE based on 3-D elasticity theory and solved by the finite element method. Formulas to extract the material constants from solutions for the RVE under three loading cases are established using the elasticity. An extended rule of mixtures, which can be used to estimate the Young's modulus in the axial direction of the RVE and to validate the numerical solutions for short CNTs, is also derived using the strength of materials theory. Numerical examples using the FEM to evaluate the effective material constants of CNT-based composites are presented, which demonstrate that the reinforcing capabilities of the CNTs in a matrix are significant. With only about 2% and 5% volume fractions of the CNTs in a matrix, the stiffness of the composite in the CNT axial direction can increase as many as 0.7 and 9.7 times for the cases of short and long CNT fibers, respectively. These simulation results, which are believed to be the first of its kind for CNT-based composites, are consistent with the experimental results reported in the literature Schadler et al. [57], Wagner et al. [58]; Qian et al. [59]. The developed extended rule of mixtures is also found to be quite effective in evaluating the stiffness of the CNT-based composites in the CNT axial direction. Many research issues need to be addressed in the modeling and simulations of CNTs in a matrix material for the development of nano-composites. Analytical methods and simulation models to extract the mechanical properties of the CNT-based nano-composites need to be further developed and verified with experimental results. The analytical method and simulation approach developed in this paper are only a preliminary study. Different type of RVEs, load cases and different solution methods should be investigated. Different interface conditions, other than perfect bonding, need to be investigated using different models to more accurately account for the interactions of the CNTs in a matrix material at the nanoscale. Nanoscale interface cracks can be analyzed using simulations to investigate the failure mechanism in nanomaterials. Interactions among a large number of CNTs in a matrix can be simulated if the computing power is available. Single-walled and multiwalled CNTs as reinforcing fibers in a matrix can be studied by simulations to find out their advantages and disadvantages. Finally, large multi scale simulation models for CNT-based composites, which can link the models at the nano, micro and macro scales, need to be developed, with the help of analytical and experimental work [56]. The three RVEs proposed in Ref. [60] and shown in Fig. 6.3 are relatively simple regarding the models and scales and pictures in Fig. 6.4 are three loading cases for the cylindrical RVE. However, this is only the first step toward more sophisticated and large-scale simulations of CNT-based composites. As the computing power and confidence in simulations of CNT-based composites increase, large scale 3-D models containing hundreds or even more CNTs, behaving linearly or nonlinearly, with coatings or of different sizes, distributed evenly or randomly, can be employed to investigate the interactions among the CNTs in a matrix and to evaluate the effective material properties. Other numerical methods can also be attempted for the modeling and simulations of CNT-based composites, which may offer some advantages over the FEM approach. For example, the boundary element method, Liu et al. [60]; Chen and Liu [61], accelerated with the fast multi pole techniques, Fu et al. [62]; Nishimura et al. [63], and the mesh free methods (Qian et al. [64])may enable one to model an RVE with thousands of CNTs in a matrix on a desktop computer. Analysis of the CNT-based composites using the boundary element method is already underway and will be reported subsequently.

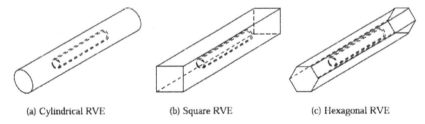

(a) Cylindrical RVE   (b) Square RVE   (c) Hexagonal RVE

FIGURE 6.3   Three nanoscale representative volume elements for the analysis of CNT-based nano-composites [56].

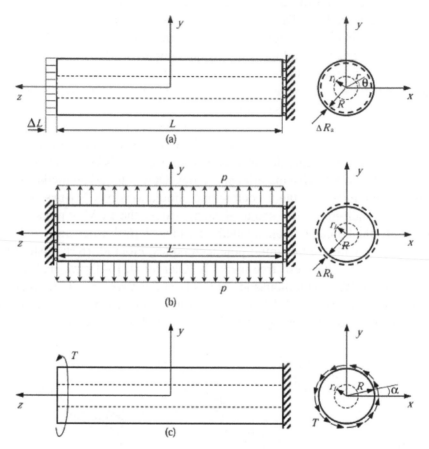

FIGURE 6.4   Three loading cases for the cylindrical RVE used to evaluate the effective material properties of the CNT-based composites. (a) Under axial stretch DL; (b) under lateral uniform load P; (c) under torsion load T [56].

The effective mechanical properties of CNT based composites are evaluated using square RVEs based on 3-D elasticity theory and solved by the FEM. Formulas to extract the effective material constants from solutions for the square RVEs under two loading cases are established based on elasticity. Square RVEs with multiple CNTs are also investigated in evaluating the Young's modulus and Poisson's ratios in the transverse plane. Numerical examples using the FEM are presented, which demonstrate that the load-carrying capabilities of the CNTs in a matrix are significant. With the addition of only about 3.6% volume fraction of the

CNTs in a matrix, the stiffness of the composite in the CNT axial direction can increase as much as 33% for the case of long CNT fibers [65]. These simulation results are consistent with both the experimental ones reported in the literature [56–59, 66]. It is also found that cylindrical RVEs tend to overestimate the effective Young's moduli due to the fact that they overestimate the volume fractions of the CNTs in a matrix. The square RVEs, although more demanding in modeling and computing, may be the preferred model in future simulations for estimating the effective material constants, especially when multiple CNTs need to be considered. Finally, the rules of mixtures, for both long and short CNT cases, are found to be quite accurate in estimating the effective Young's moduli in the CNT axial direction. This may suggest that 3-D FEM modeling may not be necessary in obtaining the effective material constants in the CNT direction, as in the studies of the conventional fiber reinforced composites. Efforts in comparing the results presented in this paper using the continuum approach directly with the MD simulations are underway. This is feasible now only for a smaller RVE of one CNT embedded in a matrix. In future research, the MD and continuum approach should be integrated in a multi scale modeling and simulation environment for analyzing the CNT-based composites. More efficient models of the CNTs in a matrix also need to be developed, so that a large number of CNTs, in different shapes and forms (curved or twisted), or randomly distributed in a matrix, can be modeled. The ultimate validation of the simulation results should be done with the nanoscale or micro scale experiments on the CNT reinforced composites [64].

Griebel and Hamaekers [67] reviewed the basic tools used in computational nano-mechanics and materials, including the relevant underlying principles and concepts. These tools range from subatomic AB initio methods to classical molecular dynamics and multiple-scale approaches. The energetic link between the quantum mechanical and classical systems has been discussed, and limitations of the standing alone molecular dynamics simulations have been shown on a series of illustrative examples. The need for multiscale simulation methods to take nanoscale aspects of material behavior was therefore emphasized; that was followed by a review and classification of the mainstream and emerging multiscale methods. These simulation methods include the broad areas of quantum mechanics, molecular dynamics and multiple-scale approaches, based on coupling the atomistic and continuum models. They summarize the strengths and limitations of currently available multiple-scale techniques, where the emphasis is made on the latest perspective approaches, such as the bridging scale method, multiscale boundary conditions, and multiscale fluidics. Example problems, in which multiple-scale simulation methods yield equivalent results to full atomistic simulations at fractions of the computational cost, were shown. They compare their results with Odegard, et al. [48], the micro mechanic method was BEM Halpin-Tsai Eq. [68] with aligned fiber by perfect bonding.

The solutions of the strain-energy-changes due to a SWNT embedded in an infinite matrix with imperfect fiber bonding are obtained through numerical method by Wan, *et al.* [69]. A "critical" SWNT fiber length is defined for full load transfer between the SWNT and the matrix, through the evaluation of the strain-energy-changes for different fiber lengths The strain-energy-change is also used to derive the effective longitudinal Young's modulus and effective bulk modulus of the composite, using a dilute solution. The main goal of their research was investigation of strain-energy-change due to inclusion of SWNT using FEM. To achieve full load transfer between the SWNT and the matrix, the length of SWNT fibers should be longer than a 'critical' length if no weak inter phase exists between the SWNT and the matrix [69].

A hybrid atomistic/continuum mechanics method is established in the Feng *et al.'s* study [70] the deformation and fracture behaviors of carbon nanotubes (CNTs) in composites. The unit cell containing a CNT embedded in a matrix is divided in three regions, which are simulated by the atomic-potential method, the continuum method based on the modified Cauchy–Born rule, and the classical continuum mechanics, respectively. The effect of CNT interaction is taken into account via the Mori–Tanaka effective field method of micromechanics. This method not only can predict the formation of Stone–Wales (5–7–7–5) defects, but also simulate the subsequent deformation and fracture process of CNTs. It is found that the critical strain of defect nucleation in a CNT is sensitive to its chiral angle but not to its diameter. The critical strain of Stone-Wales defect formation of zigzag CNTs is nearly twice that of armchair CNTs. Due to the constraint effect of matrix, the CNTs embedded in a composite are easier to fracture in comparison with those not embedded. With the increase in the Young's modulus of the matrix, the critical breaking strain of CNTs decreases.

Estimation of effective elastic moduli of nano-composites was performed by the version of effective field method developed in the framework of quasi-crystalline approximation when the spatial correlations of inclusion location take particular ellipsoidal forms [71]. The independent justified choice of shapes of inclusions and correlation holes provide the formulae of effective moduli which are symmetric, completely explicit and easily to use. The parametric numerical analyzes revealed the most sensitive parameters influencing the effective moduli which are defined by the axial elastic moduli of nano-fibers rather than their transversal moduli as well as by the justified choice of correlation holes, concentration and prescribed random orientation of nano-fibers [72].

Li and Chou [73, 74] have reported a multi scale modeling of the compressive behavior of carbon nanotube/polymer composites. The nanotube is modeled at the atomistic scale, and the matrix deformation is analyzed by the continuum finite element method. The nanotube and polymer matrix are assumed to be bonded by Vander Waals interactions at the interface. The stress distributions at the nanotube/polymer interface under isostrain and isostress loading conditions have been examined they have used beam elements for SWCNT using molecular structural mechanics, truss rod for vdW links and cubic elements for matrix, the rule of mixture was used as for comparison in this research. The buckling forces of nanotube/polymer composites for different nanotube lengths and diameters are computed. The results indicate that continuous nanotubes can most effectively enhance the composite buckling resistance.

Anumandla and Gibson [75] describes an approximate, yet comprehensive, closed form micromechanics model for estimating the effective elastic modulus of carbon nanotube-reinforced composites. The model incorporates the typically observed nanotube curvature, the nanotube length, and both 1D and 3D random arrangement of the nanotubes. The analytical results obtained from the closed form micromechanics model for nanoscale representative volume elements and results from an equivalent finite element model for effective reinforcing modulus of the nanotube reveal that the reinforcing modulus is strongly dependent on the waviness, wherein, even a slight change in the nanotube curvature can induce a prominent change in the effective reinforcement provided. The micromechanics model is also seen to produce reasonable agreement with experimental data for the effective tensile modulus of composites reinforced with multiwalled nanotubes (MWNTs) and having different MWNT volume fractions.

Effective elastic properties for carbon nanotube reinforced composites are obtained through a variety of micromechanics techniques [76]. Using the in-plane elastic properties of graphene, the effective properties of carbon nanotubes are calculated using a composite cylinders micromechanics technique as a first step in a two-step process. These effective properties are then used in the self-consistent and Mori–Tanaka methods to obtain effective elastic properties of composites consisting of aligned single or multiwalled carbon nanotubes embedded in a polymer matrix. Effective composite properties from these averaging methods are compared to a direct composite cylinders approach extended from the work of Hashin and Rosen [77], and Christensen and Lo [78]. Comparisons with finite element simulations are also performed. The effects of an inter phase layer between the nanotubes and the polymer matrix as result of fictionalization is also investigated using a multilayer composite cylinders approach. Finally, the modeling of the clustering of nanotubes into bundles due to inter atomic forces is accomplished herein using a tessellation method in conjunction with a multiphase Mori–Tanaka technique. In addition to aligned nanotube composites, modeling of the effective elastic properties of randomly dispersed nanotubes into a matrix is performed using the Mori–Tanaka method, and comparisons with experimental data are made. Selmi, et al. [79] deal with the prediction of the elastic properties of polymer composites reinforced with single walled carbon nanotubes. Our contribution is the investigation of several micromechanical models, while most of the papers on the subject deal with only one approach. They implemented four homogenization schemes, a sequential one and three others based on various extensions of the Mori–Tanaka (M–T) mean-field homogenization model: two-level (M–T/M–T), two-step (M–T/M–T) and two-step (M–T/Voigt). Several composite systems are studied, with various properties of the matrix and the graphene, short or long nanotubes, fully aligned or randomly oriented in 3D or 2D. Validation targets are experimental data or finite element results, either based on a 2D periodic unit cell or a 3D representative volume element. The comparative study showed that there are cases where all micromechanical models give adequate predictions, while for some composite materials and some properties, certain models fail in a rather spectacular fashion. It was found that the two-level (M–T/M–T) homogenization model gives the best predictions in most cases. After the characterization of the discrete nanotube structure using a homogenization method based on energy equivalence, the sequential, the two-step

(M–T/M–T), the two-step (M–T/Voigt), the two-level (M–T/M–T) and finite element models were used to predict the elastic properties of SWNT/ polymer composites. The data delivered by the micromechanical models are compared against those obtained by finite element analyzes or experiments. For fully aligned, long nanotube polymer composite, it is the sequential and the two-level (M–T/M–T) models, which delivered good predictions. For all composite morphologies (fully aligned, two-dimensional in-plane random orientation, and three-dimensional random orientation), it is the two-level (M–T/M–T) model, which gave good predictions compared to finite element and experimental results in most situations. There are cases where other micromechanical models failed in a spectacular way.

Luo, et al. [80] have used multiscale homogenization (MH) and FEM for wavy and straight SWCNTs, they have compare their results with Mori-Tanaka, Cox, Halpin-Tsai, Fu, et al. [81], Lauke [82]. Trespass, et al. [83] used 3D elastic beam for C-C bond and 3D space frame for CNT and progressive fracture model for prediction of elastic modulus, they used rule of mixture for compression of their results. Their assumption was embedded a single SWCNT in polymer with Perfect bonding. The multiscale modeling, Monte Carlo, FEM and using equivalent continuum method was used by Spanos and Kontsos [84] and compared with Zhu, et al. [85] and Paiva, et al. [86] results.

The effective modulus of CNT/PP composites is evaluated using FEA of a 3D RVE which includes the PP matrix; multiple CNTs and CNT/ PP inter phase and accounts for poor dispersion and nonhomogeneous distribution of CNTs within the polymer matrix, weak CNT/polymer interactions, CNT agglomerates of various sizes and CNTs orientation and waviness [87]. Currently, there is no other model, theoretical or numerical, that accounts for all these experimentally observed phenomena and captures their individual and combined effect on the effective modulus of nano-composites. The model is developed using input obtained from experiments and validated against experimental data. CNT reinforced PP composites manufactured by extrusion and injection molding are characterized in terms of tensile modulus, thickness and stiffness of CNT/PP inter phase, size of CNT agglomerates and CNT distribution using tensile testing, AFM and SEM, respectively. It is concluded that CNT agglomeration and waviness are the two dominant factors that hinder the great potential of CNTs as polymer reinforcement. The proposed model provides the upper and lower limit of the modulus of the CNT/PP composites and can be used to guide the manufacturing of composites with engineered properties for targeted applications. CNT agglomeration can be avoided by employing processing techniques such as sonication of CNTs, stirring, calendaring etc., whereas CNT waviness can be eliminated by increasing the injection pressure during molding and mainly by using CNTs with smaller aspect ratio. Increased pressure during molding can also promote the alignment of CNTs along the applied load direction. The 3D modeling capability presented in this study gives an insight on the upper and lower bound of the CNT/PP composites modulus quantitatively by accurately capturing the effect of various processing parameters. It is observed that when all the experimentally observed factors are considered together in the FEA the modulus prediction is in good agreement with the modulus obtained from the experiment. Therefore, it can be concluded that the FEM models proposed in this study by systematically incorporating experimentally observed characteristics can be effectively used for the determination of mechanical properties of nanocomposite materials. Their result is in agreement with the results reported in Ref. [88], The theoretical micromechanical models, shown in Fig. 6.5, are used to confirm that our FEM model predictions follow the same trend with the one predicted by the models as expected.

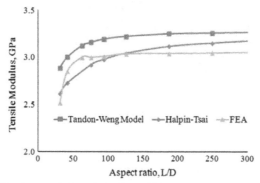

FIGURE 6.5   Effective modulus of 5wt% CNT/PP composites: theoretical models vs. FEA.

For reasons of simplicity and in order to minimize the mesh dependency on the results the hollow CNTs are considered as solid cylinders of circular cross-sectional area with an equivalent average diameter, shown in Fig. 6.6, calculated by equating the volume of the hollow CNT to the solid one [87, 88].

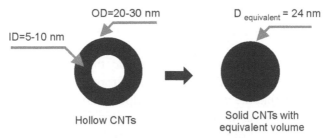

FIGURE 6.6   Schematic of the CNTs considered for the FEA.

The micromechanical models used for the comparison was Halpin–Tsai (H–T) [89] and Tandon–Weng (T–W) [90] model and the comparison was performed for 5wt% CNT/PP. It was noted that the H–T model results to lower modulus compared to FEA because H–T equation does not account for maximum packing fraction and the arrangement of the reinforcement in the composite. A modified H–T model that account for this has been proposed in the literature [91]. The effect of maximum packing fraction and the arrangement of the reinforcement within the composite become less significant at higher aspect ratios [92].

A finite element model of carbon nanotube, interphase and its surrounding polymer is constructed to study the tensile behavior of embedded short carbon nanotubes in polymer matrix in presence of vdW interactions in interphase region by Shokrieh and Rafiee [93]. The interphase is modeled using nonlinear spring elements capturing the force-distance curve of vdW interactions. The constructed model is subjected to tensile loading to extract longitudinal Young's modulus. The obtained results of this work have been compared with the results of previous research of the same authors [94] on long embedded carbon nanotube in polymer matrix. It shows that the capped short carbon nanotubes reinforce polymer matrix less efficient than long CNTs.

Despite the fact that researches have succeeded to grow the length of CNTs up to 4 cm as a world record in US Department of Energy Los Alamos National Laboratory [95] and also there are some evidences on producing CNTs with lengths up to millimeters [96, 97], CNTs are commercially available in different lengths ranging from 100 nm to approximately 30lm in the market based on employed process of growth [98–101]. Chemists at Rice University have identified a chemical process to cut CNTs into short segments [102]. As a consequent, it can be concluded that the SWCNTs with lengths smaller than 1000 nm do not contribute significantly in reinforcing polymer matrix. On the other hand, the efficient length of reinforcement for a CNT with (10, 10) index is about 1.2 lm and short CNT with length of 10.8lm can play the same role as long CNT reflecting the uppermost value reported in our previous research [94]. Finally, it is shown that the direct use of Halpin–Tsai equation to predict the modulus of SWCNT/composites overestimates the results. It is also observed that application of previously developed long equivalent fiber stiffness [94] is a good candidate to be used in Halpin–Tsai equations instead of Young's modulus of CNT. Halpin–Tsai equation is not an appropriate model for smaller lengths, since there is not any reinforcement at all for very small lengths.

Earlier, a nano-mechanical model has been developed by Chowdhury et al. [103] to calculate the tensile modulus and the tensile strength of randomly oriented short carbon nanotubes (CNTs) reinforced nano-composites, considering the statistical variations of diameter and length of the CNTs. According to this model, the entire composite is divided into several composite segments which contain CNTs of almost the same diameter and length. The tensile modulus and tensile strength of the composite are then calculated by the weighted sum of the corresponding modulus and strength of each composite segment. The existing micromechanical approach for modeling the short fiber composites is modified to account for the structure of the CNTs, to calculate the modulus and the strength of each segmented CNT reinforced composites. Multi-walled CNTs with and without intertube bridging (see Fig. 6.7) have been considered. Statistical variations of the diameter and length of the CNTs are modeled by a normal distribution. Simulation results show that CNTs intertube bridging length and diameter affect the nano-composites modulus and strength. Simulation results have been compared with the available experimental results and the comparison concludes that the developed model can be effectively used to predict tensile modulus and tensile strength of CNTs reinforced composites.

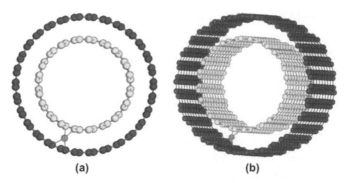

(a) (b)

FIGURE 6.7 Schematic of MWNT with intertube bridging. (a) Top view and (b) oblique view [103].

The effective elastic properties of carbon nanotube-reinforced polymers have been evaluated by Tserpes and Chanteli [104] as functions of material and geometrical parameters using a homogenized RVE. The RVE consists of the polymer matrix, a MWCNT embedded into the matrix and the interface between them. The parameters considered are the nanotube aspect ratio, the nanotube volume fraction as well as the interface stiffness and thickness For the MWCNT, both isotropic and orthotropic material properties have been considered. Analyzes have been performed by means of a 3D FE model of the RVE. The results indicate a significant effect of nanotube volume fraction. The effect of nanotube aspect ratio appears mainly at low values and diminishes after the value of 20. The interface mostly affects the effective elastic properties at the transverse direction. Having evaluated the effective elastic properties of the MWCNT-polymer at the microscale, the RVE has been used to predict the tensile modulus of a polystyrene specimen reinforced by randomly aligned MWCNTs for which experimental data exist in the literature. A very good agreement is obtained between the predicted and experimental tensile moduli of the specimen. The effect of nanotube alignment on the specimen's tensile modulus has been also examined and found to be significant since as misalignment increases the effective tensile modulus decreases radically. The proposed model can be used for the virtual design and optimization of CNT-polymer composites since it has proven capable of assessing the effects of different material and geometrical parameters on the elastic properties of the composite and predicting the tensile modulus of CNT-reinforced polymer specimens.

6.5 MODELING OF THE INTERFACE

6.5.1 INTRODUCTION

The superior mechanical properties of the nanotubes alone do not ensure mechanically superior composites because the composite properties are strongly influenced by the mechanics that govern the nanotube–polymer interface. Typically in composites, the constituents do not dissolve or merge completely and therefore, normally, exhibit an interface between one another, which can be considered as a different material with different mechanical properties. The structural strength characteristics of composites greatly depend on the nature of bonding at the interface, the mechanical load transfer from the matrix (polymer) to the nanotube and the yielding of the interface. As an example, if the composite is subjected to tensile loading and there exists perfect bonding between the nanotube and polymer and/or a strong interface then the load (stress) is transferred to the nanotube; since the tensile strength of the nanotube(or the interface) is very high the composite can withstand high loads. However, if the interface is weak or the bonding is poor, on application of high loading either the interface fails or the load is not transferred to the nanotube and the polymer fails due to their lower tensile strengths. Consider another example of transverse crack propagation. When the crack reaches the interface, it will tend to propagate along the interface, since the interface is relatively weaker (generally) than the nanotube(with respect to resistance to crack propagation). If the interface is weak, the crack will cause the interface to fracture and result in failure of the composite. In this aspect, carbon nanotubes are better than traditional fibers (glass, carbon) due to their ability to inhibit nano-and micro cracks. Hence, the knowledge and understanding of the nature and mechanics of load (stress) transfer between the nanotube and polymer and properties of the interface is critical for manufacturing of mechanically enhanced CNT-polymer composites and will enable in tailoring of the interface for specific applications or superior mechanical properties. Broadly, the interfacial mechanics of CNT-polymer composites is appealing from three aspects: mechanics, chemistry, and physics. From a mechanics point of view, the important questions are:
1.   the relationship between the mechanical properties of individual constituents, that is, nanotube and polymer, and the properties of the interface and the composite overall.

2. the effect of the unique length scale and structure of the nanotube on the property and behavior of the interface.
3. ability of the mechanics modeling to estimate the properties of the composites for the design process for structural applications.

From a chemistry point of view, the interesting issues are:

1. the chemistry of the bonding between polymer and nanotubes, especially the nature of bonding (e.g., covalent or noncovalent and electrostatic).
2. the relationship between the composite processing and fabrication conditions and the resulting chemistry of the interface.
3. the effect of functionalization (treatment of the polymer with special molecular groups like hydroxyl or halogens) on the nature and strength of the bonding at the interface.

From the physics point of view, researchers are interested in:

1. the CNT-polymer interface serves as a model nano-mechanical or a lower dimensional system (1D) and physicists are interested in the nature of forces dominating at the nano-scale and the effect of surface forces (which are expected to be significant due to the large surface to volume ratio).
2. the length scale effects on the interface and the differences between the phenomena of mechanics at the macro (or meso) and the nano-scale.

6.5.2 SOME MODELING METHOD IN INTERFACE MODELING

Computational techniques have extensively been used to study the interfacial mechanics and nature of bonding in CNT-polymer composites. The computational studies can be broadly classified as atomistic simulations and continuum methods. The atomistic simulations are primarily based on MD simulations and DFT [105, 106–110] (some references). The main focus of these techniques was to understand and study the effect of bonding between the polymer and nanotube (covalent, electrostatic or Vander Waals forces) and the effect of friction on the interface. The continuum methods extend the continuum theories of micromechanics modeling and fiber-reinforced composites (elaborated in the next section) to CNT-polymer composites [111–114] and explain the behavior of the composite from a mechanics point of view.

On the experimental side, the main types of studies that can be found in literature are as follows:

1. Researchers have performed experiments on CNT-polymer bulk composites at the macroscale and observed the enhancements in mechanical properties (like elastic modulus and tensile strength) and tried to correlate the experimental results and phenomena with continuum theories like micromechanics of composites or Kelly Tyson shear lag model [105, 115–120].
2. Raman spectroscopy has been used to study the reinforcement provided by carbon nanotubes to the polymer, by straining the CNT-polymer composite and observing the shifts in Raman peaks [121–125].
3. In situ TEM straining has also been used to understand the mechanics, fracture and failure processes of the interface. In these techniques, the CNT-polymer composite (an electron transparent thin specimen) is strained inside a TEM and simultaneously imaged to get real-time and spatially resolved (1 nm) information [110, 126].

6.5.3 NUMERICAL APPROACH

A MD model may serve as a useful guide, but its relevance for a covalent-bonded system of only a few atoms in diameter is far from obvious. Because of this, the phenomenological multiple column models that considers the interlayer radial displacements coupled through the Vander Waals forces is used. It should also be mentioned the special features of load transfer, in tension and in compression, in MWNT-epoxy composites studied by Schadler et al. [57] who detected that load transfer in tension was poor in comparison to load transfer in compression, implying that during load transfer to MWNTs, only the outer layers are stressed in tension due to the telescopic inner wall sliding (reaching at the shear stress 0.5 MPa [127]), whereas all the layers respond in compression. It should be mentioned that NTCMs usually contain not individual, separated SWCNTs, but rather bundles of closest-packed SWCNTs [128], where the twisting of the CNTs produces the radial force component giving the rope structure more stable than wires in parallel. Without strong chemically bonding, load transfer between the CNTs and the polymer matrix mainly comes from weak electrostatic and Vander Waals interactions, as well as stress/deformation arising from mismatch in the coefficients of thermal expansion [129]. Numerous researchers [130] have attributed lower than- predicted CNT-polymer composite properties to the availability of only a weak interfacial bonding. So Frankland et al. [106] demonstrated by MD simulation that the shear strength of a polymer/nano-tube interface with only Vander Waals interactions could be increased by over an order of magnitude at the occurrence of covalent bonding for only 1% of the nanotube's carbon atoms to the polymer matrix. The recent force-field-based molecular-mechanics calculations [131] demonstrated that the binding energies and frictional forces play only a minor role in determining the strength of the interface. The key factor in forming a strong bond at the interface is having a helical conformation of the polymer around the nanotube; polymer wrapping around nanotube improves the polymer-nano-tube interfacial strength although configurationally thermodynamic considerations do not necessarily support these architectures for all polymer chains [132]. Thus, the strength of the interface may result from molecular-level entanglement of the two phases and forced long-range ordering of the polymer. To ensure the robustness of data reduction schemes that are based on continuum mechanics, a careful analysis of continuum approximations used in macromolecular models and possible limitations of these approaches at the nanoscale are additionally required that can be done by the fitting of the results obtained by the use of the proposed phenomenological interface model with the experimental data of measurement of the stress distribution in the vicinity of a nanotube.

Meguid et al. [133] investigated the interfacial properties of CNT reinforced polymer composites by simulating a nanotube pullout experiment. An atomistic description of the problem was achieved by implementing constitutive relations that are derived solely from inter atomic potentials. Specifically, they adopt the Lennard-Jones (LJ) inter atomic potential to simulate a nonbonded interface, where only the Vander Waals (vdW) interactions between the CNT and surrounding polymer matrix was assumed to exist. The effects of such parameters as the CNT embedded length, the number of vdW interactions, the thickness of the interface, the CNT diameter and the cut-off distance of the LJ potential on the interfacial shear strength (ISS) are investigated and discussed. The problem is formulated for both a generic thermo set polymer and a specific two-component epoxy based on diglycidyl ether of bisphenol A (DGEBA) and triethylene tetramine (TETA) formulation. The study further illustrated that by accounting for different CNT capping scenarios and polymer morphologies around the embedded end of the CNT, the qualitative correlation between simulation and experimental pullout profiles can be improved. Only vdW interactions were considered between the atoms in the CNT and the polymer implying a nonbonded system. The vdW interactions were simulated using the LJ potential, while the CNT was described using the Modified Morse potential. The results reveal that the ISS shows a linear dependence on the vdW interaction density and decays significantly with increasing nanotube embedded length. The thickness of the interface was also varied and our results reveal that lower interfacial thicknesses favor higher ISS. When incorporating a 2.5Ψ cut-off distance to the LJ potential, the predicted ISS shows an error of approximately 25.7% relative to a solution incorporating an infinite cut-off distance. Increasing the diameter of the CNT was found to increase the peak pullout force approximately linearly. Finally, an examination of polymeric and CNT capping conditions showed that incorporating an end cap in the simulation yielded high initial pullout peaks that better correlate with experimental findings. These findings have a direct bearing on the design and fabrication of carbon nanotube reinforced epoxy composites.

Fiber pullout tests have been well recognized as the standard method for evaluating the interfacial bonding properties of composite materials. The output of these tests is the force required to pullout the nanotube from the surrounding polymer matrix and the corresponding interfacial shear stresses involved. The problem is formulated using a RVE, which consists of the reinforcing CNT, the surrounding polymer matrix, and the CNT/polymer interface as depicted in Fig. 6.8(a, b) shows a schematic of the pullout process, where x is the pullout distance and L is the embedded length of the nanotube. The atomistic-based continuum (ABC) multi scale modeling technique is used to model the RVE. The approach adopted here extends the earlier work of Wernik and Meguid [134].

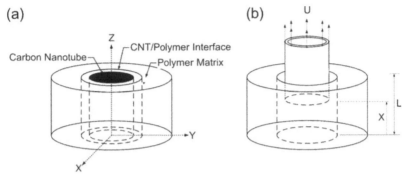

FIGURE 6.8 Schematic depictions of (a) the representative's volume element and (b) the pullout process [133].

The new features of the current work relate to the approach adopted in the modeling of the polymer matrix and the investigation of the CNT polymer interfacial properties as appose to the effective mechanical properties of the RVE. The idea behind the ABC technique is to incorporate atomistic inter atomic potentials into a continuum framework. In this way, the inter atomics potentials introduced in the model capture the underlying atomistic behavior of the different phases considered. Thus, the influence of the nano-phase is taken into account via appropriate atomistic constitutive formulations. Consequently, these measures are fundamentally different from those in the classical continuum theory. For the sake of completeness, Wernik and Meguid provided a brief outline of the method detailed in their earlier work [133, 134].

The cumulative effect of the vdW interactions acting on each CNT atom is applied as a resultant force on the respective node, which is then resolved into its three Cartesian components. This process is depicted in Fig. 6.9. During each iterations of the pullout process the above expression is reevaluated for each vdW interaction and the cumulative resultant force and its three Cartesian components are updated to correspond to the latest pullout configuration. Figure 6.10 shows a segment of the CNT with the cumulative resultant vdW force vectors as they are applied to the CNT atoms.

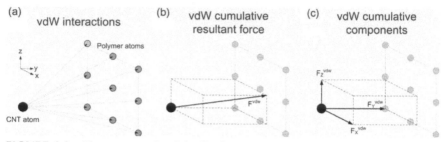

FIGURE 6.9   The process of nodal vdW force application (a) vdW interactions on an individual CNT atom, (b) the cumulative resultant vdW force, and(c) the cumulative vdW Cartesian components.

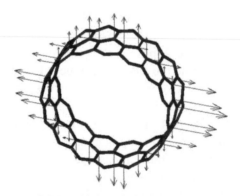

FIGURE 6.10   Segment of CNT with cumulative resultant vdW force vectors.

Yang et al. [135] investigated the CNT size effect and weakened bonding effect between an embedded CNT and surrounding matrix were characterized using MD simulations. Assuming that the equivalent continuum model of the CNT atomistic structure is a solid cylinder, the transversely isotropic elastic constants of the CNT decreased as the CNT radius increased. Regarding the elastic stiffness of the nanocomposite unit cell, the same CNT size dependency was observed in all independent components, and only the longitudinal Young's modulus showed a positive reinforcing effect whereas other elastic moduli demonstrated negative reinforcing effects as a result of poor load transfer at the interface. To describe the size effect and weakened bonding effect at the interface, a modified multi-inclusion model was derived using the concepts of an effective CNT and effective matrix. During the scale bridging process incorporating the MD simulation results and modified multi-inclusion model, we found that both the elastic modulus of the CNT and the adsorption layer near the CNT contributed to the size-dependent elastic modulus of the nano-composites. Using the proposed multi scale-bridging model, the elastic modulus for nano-composites at various volume fractions and CNT sizes could be estimated. Among three major factors (CNT waviness, the dispersion state, and adhesion between the CNT and matrix), the proposed model considered only the weakened bonding effect. However, the present multi scale framework can be easily applied in considering the aforementioned factors and describing the real nanocomposite microstructures. In addition, by considering chemically grafted molecules (covalent or noncovalent bonds) to enhance the interfacial load transfer mechanism in MD simulations, the proposed multi scale approach can offer a deeper understanding of the reinforcing mechanism, and a more practical analytical tool with which to analyze and design functional nano-composites. The analytical estimation reproduced from the proposed multi scale model can also provide useful information in modelling finite element-based representative volume elements of nanocomposite microstructures for use in multifunctional design.

The effects of the inter phase and RVE configuration on the tensile, bending and torsion properties of the suggested nanocomposite were investigated by Ayatollahi et al. [136]. It was found that the stiffness of the nanocomposite could be affected by a strong inter phase much more than by a weaker inter phase. In addition, the stiffness of the inter phase had the maximum effect on the stiffness of the nanocomposite in the bending loading conditions. Furthermore, it was revealed that the ratio of Le/Ln in RVE can dramatically affect the stiffness of the nanocomposite especially in the axial loading conditions.

For carbon nanotubes not well bonded to polymers, Jiang et al. [137] established a cohesive law for carbon nanotube/polymer interfaces. The cohesive law and its properties (e.g., cohesive strength, cohesive energy) are obtained directly from the Lennard–Jones potential from the Vander Waals interactions. Such a cohesive law is incorporated in the micromechanics model to study the mechanical behavior of carbon nanotube-reinforced composite materials. Carbon nanotubes indeed improve the mechanical behavior of composite at the small strain. However, such improvement disappears at relatively large strain because the completely deboned nanotubes behave like voids in the matrix and may even weaken the composite. The increase of interface adhesion between carbon nanotubes and polymer matrix may significantly improve the composite behavior at the large strain [138].

Zalamea et al. [139]employed the shear transfer model as well as the shear lag model to explore the stress transfer from the outermost layer to the interior layers in MWCNTs. Basically, the interlayer properties between graphene layers were designated by scaling the parameter of shear transfer efficiency with respect to the perfect bonding. Zalamea et al. pointed out that as the number of layers in MWCNTs increases, the stress transfer efficiency decreases correspondingly. Shen et al. [140] examined load transfer between adjacent walls of DWCNTs using MD simulation, indicating that the tensile loading on the outermost wall of MWCNTs cannot be effectively transferred into the inner walls. However, when chemical bonding between the walls is established, the effectiveness can be dramatically enhanced. It is noted that in the above investigations, the loadings were applied directly on the outermost layers of MWCNTs; the stresses in the inner layers were then calculated either from the continuum mechanics approach [139] or MD simulation [140]. Shokrieh and Rafiee [93, 94] examined the mechanical properties of nano-composites with capped SWCNTs embedded in a polymer matrix. The load transfer efficiency in terms of different CNTs' lengths was the main concern in their examination. By introducing an inter phase to represent the vdW interactions between SWCNTs and the surrounding matrix, Shokrieh and Rafiee [93, 94] converted the atomistic SWCNTs into an equivalent continuum fiber in finite element analysis. The idea of an equivalent solid fiber was also proposed by Gao and Li [141] to replace the atomistic structure of capped SWCNTs in the nano-composites' cylindrical unit cell. The modulus of the equivalent solid was determined based on the atomistic structure of SWCNTs through molecular structure mechanics [142]. Subsequently, the continuum-based shear lag analysis was carried out to evaluate the axial stress distribution in CNTs. In addition, the influence of end caps in SWCNTs on the stress distribution of nano-composites was also taken into account in their analysis. Tsai and Lu [143] characterized the effects of the layer number, intergraphic layers interaction, and aspect ratio of MWCNTs on the load transfer efficiency using the conventional shear lag model and finite element analysis. However, in their analysis, inter atomistic characteristics of the adjacent graphene layers associated with different degrees of interactions were simplified by a thin inter phase with different moduli. The atomistic interaction between the grapheme layers was not taken into account in their modeling of MWCNTs. In light of the forgoing investigations, the equivalent solid of SWCNTs was developed by several researchers and then implemented as reinforcement in continuum-based nanocomposite models. Nevertheless, for MWCNTs, the subjects concerning the development of equivalent continuum solid are seldom explored in the literature. In fact, how to introduce the atomistic characteristics, that is, the interfacial properties of neighboring graphene layers in MWCNTs, into the equivalent continuum solid is a challenging task as the length scales used to describe the physical phenomenon are distinct. Thus, a multiscale-based simulation is required to account for the atomistic attribute of MWCNTs into an equivalent continuum solid. In Lu and Tsai's study [144], the multiscale approach was used to investigate the load transfer efficiency from surrounding matrix to DWCNTs. The analysis consisted of two stages. First, a cylindrical DWCNTs equivalent continuum was proposed based on MD simulation where the pullout extension on the outer layer was performed in an attempt to characterize the atomistic behaviors between neighboring graphite layers. Subsequently, the cylindrical continuum (denoting the DWCNTs) was embedded in a unit cell of nano-composites, and the axial stress distribution as well as the load transfer efficiency of the DWCNTs was evaluated from finite element analysis. Both SWCNTs and DWCNTs were considered in the simulation and the results were compared with each other.

An equivalent cylindrical solid to represent the atomistic attributes of DWCNTs was proposed in this study. The atomistic interaction of adjacent graphite layers in DWCNTs was characterized using MD simulation based on which a spring element was introduced in the continuum equivalent solid to demonstrate the interfacial properties of DWCNTs. Subsequently, the proposed continuum solid (denotes DWCNTs) was embedded in the matrix to form DWCNTs nano-composites (continuum model), and the load transfer efficiency within the DWCNTs was determined from FEM analysis. For the demonstration purpose, the DWCNTs with four different lengths were considered in the investigation. Analysis results illustrate that the increment of CNTs' length can effectively improve the load transfer efficiency in the outermost layers, nevertheless, for the inner layers, the enhancement is miniature. On the other hand, when the covalent bonds between the adjacent graphene layers are crafted, the load carrying capacity in the inner layer increases as so does the load transfer efficiency of DWCNTs. As compared to SWCNTs, the DWCNTs still possess the less capacity of load transfer efficiency even though there are covalent bonds generated in the DWCNTs.

6.6   CONCLUDING REMARKS

Many traditional simulation techniques (e.g., MC, MD, BD, LB, Ginz-burg–Landau theory, micromechanics and FEM) have been employed, and some novel simulation techniques (e.g., DPD, equivalent-continuum and self-similar approaches) have been developed to study polymer na-no-composites. These techniques indeed represent approaches at various time and length scales from molecular scale (e.g., atoms), to micro scale (e.g., coarse-grains, particles, monomers) and then to macroscale (e. g, do-mains), and have shown success to various degrees in addressing many as-pects of polymer nano-composites. The simulation techniques developed thus far have different strengths and weaknesses, depending on the need of research. For example, molecular simulations can be used to investigate molecular interactions and structure on the scale of 0.1–10 nm. The result-ing information is very useful to understanding the interaction strength at nanoparticle–polymer interfaces and the molecular origin of mechanical improvement. However, molecular simulations are computationally very demanding, thus not so applicable to the prediction of mesoscopic structure and properties defined on the scale of 0.1–10 mm, for example, the disper-sion of nanoparticles in polymer matrix and the morphology of polymer nano-composites. To explore the morphology on these scales, mesoscopic simulations such as coarse-grained methods, DPD and dynamic mean field theory are more effective. On the other hand, the macroscopic properties of materials are usually studied by the use of mesoscale or macroscale techniques such as micromechanics and FEM. But these techniques may have limitations when applied to polymer nano-composites because of the difficulty to deal with the interfacial nanoparticle–polymer interaction and the morphology, which are considered crucial to the mechanical improve-ment of nanoparticle-filled polymer nano-composites. Therefore, despite the progress over the past years, there are a number of challenges in com-puter modeling and simulation. In general, these challenges represent the work in two directions. First, there is a need to develop new and improved simulation techniques at individual time and length scales. Secondly, it is important to integrate the developed methods at wider range of time and length scales, spanning from quantum mechanical domain (a few atoms) to molecular domain (many atoms), to mesoscopic domain (many monomers or chains), and finally to macroscopic domain (many domains or structures), to form a useful tool for exploring the structural, dynamic, and mechanical properties, as well as optimizing design and processing control of polymer nano-composites. The need for the second development is obvious. For example, the morphology is usually determined from the mesoscale techniques whose implementation requires information about the interactions between various components (e.g., nanoparticle–nanoparticle and nanoparticle–polymer) that should be derived from molecular simulations. Developing such a multi scale method is very challenging but indeed represents the future of computer simulation and modeling, not only in polymer nano-composites but also other fields. New concepts, theories and computational tools should be developed in the future to make truly seamless multi scale modeling a reality. Such development is crucial in order to achieve the longstanding goal of predicting particle–structure property relationships in material design and optimization.

The strength of the interface and the nature of interaction between the polymer and carbon nanotube are the most important factors governing the ability of nanotubes to improve the performance of the composite. Extensive research has been performed on studying and understanding CNT-polymer composites from chemistry, mechanics and physics aspects. However, there exist various issues like processing of composites and experimental challenges, which need to be addressed to gain further insights into the interfacial processes.

KEYWORDS

- **CNT-polymer composites**
- **Material design and optimization**
- **Mesoscopic domain**
- **Micromechanics**
- **Modeling and simulation techniques**
- **Modeling of the Interface**

REFERENCES

1. Bandosz, T. J. (2006). Activated Carbon Surfaces in Environmental Remediation, *7*, Academic Press.
2. Marsh, H., & Rodriguez-Reinoso, F. (2006). *Activated Carbon*, Elsevier Science Limited.
3. Bach, M. T. (2007). *Impact of Surfaces Chemistry on Adsorptions*, Tailoring of Activated Carbons University of Florida.
4. Machnikowski, J., et al. (2010). Tailoring Porosity Development in Monolithic Adsorbents Made of KOH-Activated Pitch Coke and Furfurals Alcohol Binder for Methane Storage, Energy & Fuels, *24(6)*, 3410–3414.
5. Yin, C. Y., Aroua, M. K., & Daud, W. M. A. W. (2007). Review of Modifications of Activated Carbon for Enhancing Contaminant uptakes from Aqueous Solutions, *Separation and Purification Technology, 52(3)*, 403–415.
6. Muniz, J., Herrero, J., & Fuertes, A. (1998). Treatments to Enhance the SO_2 Capture by Activated Carbon Fibers, *Applied Catalysis B: Environmental, 18(1)*, 171–179.
7. Houshmand, A., Daud, W. M. A. W., & Shafeeyan, M. S. (2011) Tailoring the Surface Chemistry of Activated Carbon by Nitric Acid, Study Using Response Surface Method, Bulletin of the Chemical Society of Japan, *84(11)*, 1251–1260.
8. Kyotani, T. (2000). Control of Pore Structure in Carbon, *Carbon, 38(2)*, 269–286.
9. Kaneko, K. (1994). Determination of Pore Size and Pore Size Distribution, 1. Adsorbents and Catalysts, *Journal of Membrane Science, 96(1)*, 59–89.
10. Olivier, J. P. (1998). Improving the Models Used for Calculating the Size Distribution of Micro pore Volume of Activated Carbons from Adsorption Data, *Carbon, 36(10)*, 1469–1472.
11. Lastoskie, C., Gubbins, K. E., & Quirke, N. (1993). Pore Size Heterogeneity and the Carbon Slit Pore, A Density Functional Theory Model, *Langmuir, 9(10)*, 2693–2702.
12. Cao, D., & Wu, J. (2005). Modeling the Selectivity of Activated Carbons for Efficient Separation of Hydrogen and Carbon Dioxide, *Carbon, 43(7)*, 1364–1370.
13. Feng, W. et al. (2005). Adsorption of Hydrogen Sulfide onto Activated Carbon Fibers, Effect of Pore Structure and Surface Chemistry, *Environmental Science and Technology, 39(24)*, 9744–9749.
14. Banerjee, R. et al. (2009). Control of Pore Size and Functionality in Reticular Zeolitic Imidazolate Frameworks and their Carbon Dioxide Selective Capture Properties, *Journal of the American Chemical Society, 131(11)*, 3875–3877.
15. Kowalczyk, P. et al. (2003). Estimation of the Pore Size Distribution Functions from the Nitrogen Adsorption Isotherm, Comparison of Density Functional Theory and The Method of Do and Co-workers, *Carbon, 41(6)*, 1113–1125.
16. Tseng, R. L., & Tseng, S. K. (2005). Pour Structure and Adsorption Performance of the KOH-Activated Carbons Prepared from Corncob, *Journal of Colloid and Interface Science, 287(2)*, 428–437.

17. Kowalczyk, P., Ciach, A., & Neimark, A. V. (2008). Adsorption-Induced Deformation of Micro Porous Carbons, Pore Size Distribution Effect, *Langmuir, 24(13),* 6603–6608.
18. Terzyk, A. P. et al. (2007). How Realistic is the Pore Size Distribution Calculated from Adsorption Isotherms if Activated Carbon is composed of Fullerene-like Fragments? *Physical Chemistry Chemical Physics, 9(44),* 5919–5927.
19. Mangun, C. et al. (1998). Effect of Pore Size on Adsorption of Hydrocarbons in Phenolic-based Activated Carbon Fibers, *Carbon, 36(1),* 123–129.
20. Endo, M. et al. (2001). High Power Electric Double Layer Capacitor (EDLC's), from Operating Principle to Pore Size Control in Advanced Activated Carbons, *Carbon Science, 1(3, 4),* 117–128.
21. Gadkaree, K., & Jaroniec, M. (2000). Pore Structure Development in Activated Carbon Honeycombs, *Carbon, 38(7),* 983–993.
22. Pelekani, C., & Snoeyink, V. L. (2000). Competitive Adsorption between Atrazine and Methylene Blue on Activated Carbon, the Importance of Pore Size Distribution, *Carbon, 38(10),* 1423–1436.
23. Daud, W. M. A. W., Ali, W. S. W., & Sulaiman, M. Z. (2000). The Effects of Carbonization Temperature on Pore Development in Palm-Shell-Based Activated Carbon, *Carbon, 38(14),* 1925–1932
24. Pelekani, C., & Snoeyink, V. L. (2001). A Kinetic and Equilibrium Study of Competitive Adsorption between Atrazine and Congo Red Dye on Activated Carbon, the Importance of Pore Size Distribution, *Carbon, 39(1),* 25–37.
25. Ebie, K. et al. (2001). Pore Distribution Effect of Activated Carbon in Adsorbing Organic Micro Pollutants from Natural Water, *Water Research, 35(1),* 167–179.
26. Mangun, C. L. et al. (2001). Surface Chemistry, Pore Sizes and Adsorption Properties of Activated Carbon Fibers and Precursors Treated with Ammonia, *Carbon, 39(12),* 1809–1820.
27. Moreira, R., Jose, H., & Rodrigues, A. (2001). Modification of Pore Size in Activated Carbon by Polymer Deposition and its Effects on Molecular Sieve Selectivity, *Carbon, 39(15),* 2269–2276.
28. Yang, J. B. et al. (2002). Preparation and Properties of Phenolic Resin-Based Activated Carbon Spheres with Controlled Pore Size Distribution, *Carbon, 40(6),* 911–916.
29. Cao, D. et al. (2002). Determination of Pore Size Distribution and Adsorption of Methane and CCl_4 on Activated Carbon by Molecular Simulation, *Carbon, 40(13),* 2359–2365.
30. Py, X., Guillot, A., & Cagnon, B. (2003). Activated Carbon Porosity Tailoring by Cyclic Sorption/Decomposition of Molecular Oxygen, *Carbon, 41(8),* 1533–1543.
31. Tanaike, O. et al. (2003). Preparation and Pore Control of Highly Mesoporous Carbon from Defluorinated PTFE. *Carbon, 41(9),* 1759–1764.
32. Zhao, J. et al. (2007). Pore Structure Control of Mesoporous Carbon as Super Capacitor Material. *Materials Letters, 61(23),* 4639–4642.
33. Chmiola, J. et al. (2006). Anomalous Increase in Carbon Capacitance at Pore Sizes Less than 1 Nanometer, *Science, 313(5794),* 1760–1763.
34. Lin, C., Ritter, J. A., & Popov, B. N. (1999). Correlation of Double-Layer Capacitance with the Pore Structure of Sol-Gel Derived Carbon Xerogels, *Journal of the Electrochemical Society, 146(10),* 3639–3643.

35. Gogotsi, Y. et al. (2003). Nano Porous Carbide-Derived Carbon with Tunable Pore Size, *Nature Materials, 2(9)*, 591–594.
36. Ustinov, E., Do, D., & Fenelonov, V. (2006). Pore Size Distribution Analysis of Activated Carbons, Application of Density Functional Theory Using Non Graphitized Carbon Black as a Reference System, *Carbon, 44(4)*, 653–663.
37. Han, S. et al. (2003). The Effect of Silica Template Structure on the Pore Structure of Mesoporous Carbons, *Carbon, 41(5)*, 1049–1056.
38. Pérez-Mendoza, M. et al. (2006). Analysis of the Micro Porous Texture of a Glassy Carbon by Adsorption Measurements and Monte Carlo simulation, Evolution with Chemical and Physical Activation, *Carbon, 44(4)*, 638–645.
39. Dombrowski, R. J., Hyduke, D. R., & Lastoskie, C. M. (2000). Pore Size Analysis of Activated Carbons from Argon and Nitrogen Porosimetry Using Density Functional Theory, *Langmuir, 16(11)*, 5041–5050.
40. Khalili, N. R. et al. (2000). Production of Micro-and Mesoporous Activated Carbon from Paper Mill Sludge, I. Effect of Zinc Chloride Activation, *Carbon, 38(14)*, 1905–1915.
41. Dandekar, A., Baker, R., & Vannice, M. (1998). Characterization of Activated Carbon, Graphitized Carbon Fibers and Synthetic Diamond Powder Using TPD and DRIFTS, *Carbon, 36(12)*, 1821–1831.
42. Lastoskie, C., Gubbins, K. E., & Quirke, N. (1993). Pore Size Distribution Analysis of Micro Porous Carbons, a Density Functional Theory Approach, *The Journal of Physical Chemistry, 97(18)*, 4786–4796.
43. Kakei, K. et al. (1990). Multi-Stage Micro Pores Filling Mechanism of Nitrogen on Micro Porous and Micro Graphitic Carbons, *J. Chem. Soc., Faraday Trans., 86(2)*, 371–376.
44. Sing, K. S. (1998). Adsorption Methods for the Characterization of Porous Materials, *Advances in Colloid and Interface Science, 76*, 3–11.
45. Barranco, V. et al. (2010). Amorphous Carbon Nano Fibers and their Activated Carbon Nano Fibers as Super Capacitor Electrodes, *Journal of Physical Chemistry C, 114(22)*, 10302–10307.
46. Kawabuchi, Y. et al. (1997). Chemical Vapor Deposition of Heterocyclic Compounds over Active Carbon Fiber to Control its Porosity and Surface Function, *Langmuir, 13(8)*, 2314–2317.
47. Miura, K., Hayashi, J., & Hashimoto, K. (1991). Production of Molecular Sieving Carbon through Carbonization of Coal Modified by Organic Additives, *Carbon, 29(4)*, 653–660.
48. Kawabuchi, Y. et al. (1998). The Modification of Pore Size in Activated Carbon Fibers by Chemical Vapor Deposition and its Effects on Molecular Sieve Selectivity, *Carbon, 36(4)*, 377–382.
49. Verma, S. & Walker, P. (1993). Preparation of Carbon Molecular Sieves by Propylene Pyrolysis over Nickel-Impregnated Activated Carbons, *Carbon, 31(7)*, 1203–1207.
50. Verma, S., Nakayama, Y., & Walker, P. (1993). Effect of Temperature on Oxygen-Argon Separation on Carbon Molecular Sieves, *Carbon, 31(3)*, 533–534.
51. Chen, Y. & Yang, R. (1994). Preparation of Carbon Molecular Sieve Membrane and Diffusion of Binary Mixtures in the Membrane, *Industrial & Engineering Chemistry Research, 33(12)*, 3146–3153.

52. Rao, M., & Sircar, S. (1996). Performance and Pore Characterization of Nano Porous Carbon Membranes for Gas Separation, *Journal of Membrane Science, 110(1)*, 109–118.

53. Katsaros, F., et al. (1997). High Pressure Gas Permeability of Micro Porous Carbon Membranes, *Micro Porous Materials, 8(3)*, 171–176.

54. Kang, I. (2005). Carbon Nano Tube Smart Materials, University of Cincinnati.

55. Khan, Z. H., & Husain, M. (2005). Carbon Nano Tube and its Possible Applications. *Indian Journal of Engineering and Materials Sciences, 12(6)*, 529.

56. Khare, R., & Bose, S. (2005). Carbon Nano Tube Based Composites a Review, *Journal of Minerals & Materials Characterization & Engineering, 4(1)*, 31–46.

57. Abuilaiwi, F. A. et al. (2010). Modification and Functionalization of Multi Walled Carbon Nanotube (MWCNT) via Fischer Esterification, *Arabian Journal for Science and Engineering, 35(1c)*, 37–48.

58. Gupta, S., & Farmer, J. (2011). Multi Walled Carbon Nano Tubes and Dispersed Nano Diamond Novel Hybrids, Microscopic Structure Evolution, Physical Properties, and Radiation Resilience, *Journal of Applied Physics, 109(1)*, 014314.

59. Upadhyayula, V. K., & Gadhamshetty, V. (2010). Appreciating the Role of Carbon Nano Tube Composites in Preventing Bio Fouling and Promoting Bio Films on Material Surfaces in Environmental Engineering: A review. *Biotechnology advances, 28(6)*, 802–816.

60. Saba, J. et al. (2012). Continuous Electro Deposition of Polypyrrole on Carbon Nano Tube-Carbon Fiber Hybrids as a Protective Treatment against Nanotube Dispersion, *Carbon.*

61. Kim, W. D. et al. (2011). Tailoring the Carbon Nanostructures Grown on the Surface of Ni–Al Bimetallic Nanoparticles in the Gas Phase, *Journal of Colloid and Interface Science, 362(2)*, 261–266.

62. Schwandt, C., Dimitrov, A., & Fray, D. (2010). The Preparation of Nano-Structured Carbon Materials by Electrolysis of Molten Lithium Chloride at Graphite Electrodes, *Journal of Electro Analytical Chemistry, 647(2)*, 150–158.

63. Gao, C. et al. (2012). The New Age of Carbon Nano Tubes, an Updated Review of Functionalized Carbon Nano Tubes in Electrochemical Sensors. *Nanoscale, 4(6)*, 1948–1963.

64. Ben-Valid, S. et al. (2010). Spectroscopic and Electrochemical Study of Hybrids Containing Conductive Polymers and Carbon Nano Tubes, *Carbon, 48(10)*, 2773–2781.

65. Vecitis, C. D., Gao, G., & Liu, H. (2011). Electrochemical Carbon Nano Tube Filter for Adsorption, Desorption, and Oxidation of Aqueous Dyes and Anions, *the Journal of Physical Chemistry C, 115(9)*, 3621–3629.

66. Jagannathan, S. et al. (2008). Structure and Electrochemical Properties of Activated Polyacrylonitrile Based Carbon Fibers Containing Carbon Nano Tubes, *Journal of Power Sources, 185(2)*, 676–684.

67. Zhu, Y. et al. (2011). Carbon-based Super Capacitors Produced by Activation of Graphene, *Science, 332(6037)*, 1537–1541.

68. Obreja, V. V. (2008). On the Performance of Super Capacitors with Electrodes Based on Carbon Nano Tubes and Carbon Activated Material-a Review Physical E, *Low-dimensional Systems and Nanostructures, 40(7)*, 2596–2605.

69. Wang, L., & Yang, R. T. (2010). Hydrogen Storage on Carbon-Based Adsorbents and Storage at Ambient Temperature by Hydrogen Spillover. *Catalysis Reviews Science and Engineering, 52(4)*, 411–461.
70. Dillon, A. et al. (1997). Carbon Nanotube Materials for Hydrogen Storage, Proceedings of the DOE/NREL Hydrogen Program Review, *237*.
71. Ströbel, R. et al. (2006). Hydrogen Storage by Carbon Materials, *Journal of Power Sources, 159(2)*, 781–801.
72. Gupta, V. K., & Saleh, T. A. (2013). Sorption of Pollutants by Porous Carbon, Carbon Nano Tubes and Fullerene: An Overview. *Environ Sci. Pollut. Res., 20*, 2828–2843.
73. Gupta, V. K. et al. (2013). Adsorptive Removal of Dyes from Aqueous Solution onto Carbon Nano Tubes: A Review, *Advances in Colloid and Interface Science, 193–194*, 24–34.
74. Yang, K., et al. (2006). Competitive Sorption of Pyrene, Phenanthrene and Naphthalene on Multi Walled Carbon Nano Tubes, *Environmental Science and Technology, 40(18)*, 5804–5810.
75. Zhang, H. et al. (2011). Synthesis of a Novel Composite Imprinted Material Based on Multi Walled Carbon Nano Tubes as a Selective Melamine Absorbent, *Journal of Agricultural and Food Chemistry, 59(4)*, 1063–1071.
76. Yan, L., et al. (2012). Characterization of Magnetic Guar Gum-Grafted Carbon Nano Tubes and the Adsorption of the Dyes, *Carbohydrate Polymers, 87(3)*, 1919–1924.
77. Wu, C. H. (2007). Adsorption of Reactive Dye onto Carbon Nano Tubes, Equilibrium, Kinetics and Thermodynamics, *Journal of Hazardous Materials, 144(1–2)*, 93–100.
78. Vadi, M., & Ghaseminejhad, E. (2011). Comparative Study of Isotherms Adsorption of Oleic Acid by Activated Carbon and Multi-wall Carbon Nano Tube, *Oriental Journal of Chemistry, 27(3)*, 973.
79. Vadi, M., & Moradi, N. (2011). Study of Adsorption Isotherms of Acetamide and Propionamide on Carbon Nano Tube, *Oriental Journal of Chemistry, 27(4)*, 1491.
80. Mishra, A. K., Arockiadoss, T., & Ramaprabhu, S. (2010). Study of Removal of Azo Dye by Functionalized Multi Walled Carbon Nano Tubes, *Chemical Engineering Journal, 162(3)*, 1026–1034.
81. Madrakian, T. et al. (2011). Removal of Some Cationic Dyes from Aqueous Solutions Using Magnetic-Modified Multi-Walled Carbon Nano Tubes, *Journal of Hazardous Materials, 196*, 109–114.
82. Kuo, C. Y., Wu, C. H., & Wu, J. Y. (2008). Adsorption of Direct Dyes from Aqueous Solutions by Carbon Nano Tubes, Determination of Equilibrium, Kinetics and Thermodynamics Parameters, *Journal of Colloid and Interface Science, 327(2)*, 308–315.
83. Gong, J. L. et al. (2009). Removal of Cationic Dyes from Aqueous Solution using Magnetic Multi-wall Carbon Nano Tube Nanocomposite as Adsorbent, *Journal of Hazardous Materials, 164(2)*, 1517–1522.
84. Chang, P. R. et al. (2011). Characterization of Magnetic Soluble Starch-Functionalized Carbon Nano Tubes and its Application for the Adsorption of the Dyes, *Journal of Hazardous Materials, 186(2)*, 2144–2150.
85. Chatterjee, S., Lee, M. W., & Woo, S. H. (2010). Adsorption of Congo red by Chitosan Hydrogel Beads Impregnated with Carbon Nano Tubes, *Bio resource Technology, 101(6)*, 1800–1806.

86. Chen, Z. et al. (2011). Adsorption Behavior of Epirubicin Hydrochloride on Carboxylated Carbon Nano Tubes, *International Journal of Pharmaceutics, 405(1),* 153–161.
87. Ai, L. et al. (2011), Removal of methylene blue from aqueous solution with magnetite loaded multiwall carbon Nano Tube, Kinetic, Isotherm and Mechanism Analysis, *Journal of Hazardous Materials, 198,* 282–290.
88. Chronakis, I. S. (2005). Novel Nano Composites and Nano Ceramics Based on Polymer Nano Fibers Using Electro Spinning Process, *Review Journal of Materials Processing Technology, 167(2),* 283–293.
89. Inagaki, M., Yang, Y., & Kang, F. (2012). Carbon nano Fibers Prepared via Electro Spinning, *Advanced Materials, 24(19),* 2547–2566.
90. Im, J. S. et al. (2008). The Study of Controlling Pore Size on Electro Spun Carbon Nano Fibers for Hydrogen Adsorption, *Journal of Colloid and Interface Science, 318(1),* 42–49.
91. De Jong, K. P., & Geus, J. W. (2000). Carbon Nano Fibers, Catalytic Synthesis and Applications, *Catalysis Reviews, 42(4),* 481–510.
92. Huang, J., Liu, Y., & You, T. (2010). Carbon Nano Fiber Based Electrochemical Biosensors, *A Review Analytical Methods, 2(3),* 202–211.
93. Yusof, N., & Ismail, A. (2012). Post Spinning and Pyrolysis Processes of Polyacrylonitrile (PAN)-Based Carbon Fiber and Activated Carbon Fiber, *A Review Journal of Analytical and Applied Pyrolysis, 93,* 1–13.
94. Sullivan, P. et al. (2012). Physical and Chemical Properties of PAN-derived Electro Spun Activated Carbon Nano Fibers and their Potential for Use as an Adsorbent for Toxic Industrial Chemicals, *Adsorption, 18(3–4),* 265–274.
95. Tavanai, H., Jalili, R., & Morshed, M. (2009). Effects of Fiber Diameter and CO_2 Activation Temperature on the Pore Characteristics of Polyacrylonitrile Based Activated Carbon Nano Fibers, *Surface and Interface Analysis, 41(10),* 814–819.
96. Wang, G. et al. (2012). Activated Carbon Nano Fiber Webs Made by Electro Spinning for Capacitive Deionization, *Electro Chemical Acta, 69,* 65–70.
97. Ra, E. J. et al. (2010). Ultra Micro Pore Formation in PAN/camphor-based Carbon Nano Fiber Paper, *Chemical Communications, 46(8),* 1320–1322.
98. Liu, W., & Adanur, S. (2010). Properties of Electro Spun Polyacrylonitrile Membranes and Chemically Activated Carbon Nano Fibers, *Textile Research Journal, 80(2),* 124–134.
99. Korovchenko, P., Renken, A., & Kiwi-Minsker, L. (2005). Microwave Plasma Assisted Preparation of Pd-nanoparticles with Controlled Dispersion on Woven Activated Carbon Fibers, *Catalysis Today, 102,* 133–141.
100. Jung, K. H., & Ferraris, J. P. (2012). Preparation and Electrochemical Properties of Carbon Nano Fibers Derived from Polybenzimidazole/Polyimide Precursor Blends, *Carbon.*
101. Im, J. S., Park, S. J., & Lee, Y. S. (2007). Preparation and Characteristics of Electro Spun Activated Carbon Materials having Meso-and Macro Pores, *Journal of Colloid and Interface Science, 314(1),* 32–37.
102. Esrafilzadeh, D., Morshed, M., & Tavanai, H. (2009). An Investigation on the Stabilization of Special Polyacrylonitrile Nano Fibers as Carbon or Activated Carbon Nano Fiber Precursor, *Synthetic Metals, 159(3),* 267–272.

103. Hung, C. M. (2009). Activity of Cu-activated Carbon Fiber Catalyst in Wet Oxidation of Ammonia Solution, *Journal of Hazardous Materials, 166(2)*, 1314–1320.
104. Koslow, E. E. (2007). Carbon or Activated Carbon Nano Fibers, *Google Patents*.
105. Lee, J. W. et al. (2006). Heterogeneous Adsorption of Activated Carbon Nano Fibers Synthesized by Electro Spinning Polyacrylonitrile Solution, *Journal of Nano Science and Nanotechnology, 6(11)*, 3577–3582.
106. Oh, G. Y. et al. (2008). Preparation of the Novel Manganese-Embedded PAN-Based Activated Carbon Nano Fibers by Electro Spinning and their Toluene Adsorption, *Journal of Analytical and Applied Pyrolysis, 81(2)*, 211–217.
107. Zussman, E. et al. (2005). Mechanical and Structural Characterization of Electro Spun PAN-derived Carbon Nano Fibers, *Carbon, 43(10)*, 2175–2185.
108. Kim, C. (2005). Electrochemical Characterization of Electro Spun Activated Carbon Nano Fibers as an Electrode in Super Capacitors, *Journal of Power Sources, 142(1)*, 382–388.
109. Jung, M. J. et al. (2013). Influence of the Textual Properties of Activated Carbon Nano Fibers on the Performance of Electric Double-Layer Capacitors, *Journal of Industrial and Engineering Chemistry*.
110. Fan, Z. et al. (2011). Asymmetric Super Capacitors Based on Graphene/MnO$_2$ and Activated Carbon Nano Fiber Electrodes with High Power and Energy Density, *Advanced Functional Materials, 21(12)*, 2366–2375.
111. Jeong, E., Jung, M. J., & Lee, Y. S. (2013). Role of Fluorination in Improvement of the Electrochemical Properties of Activated Carbon Nano Fiber Electrodes, *Journal of Fluorine Chemistry*.
112. Seo, M. K., & Park, S. J. (2009). Electrochemical Characteristics of Activated Carbon Nano Fiber electrodes for Super Capacitors, *Materials Science and Engineering: B, 164(2)*, 106–111.
113. Endo, M. et al. (2001). High Power Electric Double Layer Capacitor (EDLC's), from Operating Principle to Pore Size Control in Advanced Activated Carbons, *Carbon Science, 1(3&4)*, 117–128.
114. Ji, L. & Zhang, X., (2009). Generation of Activated Carbon Nano Fibers from Electrospun Polyacrylonitrile-Zinc Chloride Composites for use as Anodes in Lithium-Ion Batteries. *Electrochemistry Communications, 11(3)*, 684–687.
115. Karra, U. et al. (2012). Power Generation and Organics Removal from Wastewater Using Activated Carbon Nano Fiber (ACNF) Microbial Fuel Cells (MFCs). *International Journal of Hydrogen Energy*.
116. Oh, G. Y. et al. (2008). Adsorption of Toluene on Carbon Nano Fibers Prepared by Electro Spinning, *Science of the Total Environment, 393(2)*, 341–347.
117. Lee, K. J. et al. (2010). Activated Carbon Nano Fiber Produced from Electro Spun Polyacrylonitrile Nano Fiber as a Highly Efficient Formaldehyde Adsorbent, *Carbon, 48(15)*, 4248–4255.
118. Katepalli, H. et al. (2011). Synthesis of Hierarchical Fabrics by Electro Spinning of PAN Nano Fibers on Activated Carbon Microfibers for Environmental Remediation Applications, *Chemical Engineering Journal, 171(3)*, 1194–1200.
119. Gaur, V., Sharma, A., & Verma, N. (2006). Preparation and Characterization of ACF for the Adsorption of BTX and SO$_2$, Chemical Engineering and Processing, *Process Intensification, 45(1)*, 1–13.

120. Cheng, K. K., Hsu, T. C., & Kao, L. H. (2011). A Microscopic View of Chemically Activated Amorphous Carbon Nano Fibers Prepared from Core/Sheath Melt-Spinning of Phenol Formaldehyde-Based Polymer Blends, *Journal of Materials Science, 46(11)*, 3914–3922.

121. Zhijun, et al. (2009). A Molecular Simulation Probing of Structure and Interaction for Supra Molecular Sodium Dodecyl Sulfate/S-W Carbon Nano Tube Assemblies, Nano Lett.

122. Ong, et al. (2010). Molecular Dynamics Simulation of Thermal Boundary Conductance between Carbon Nano Tubes and SiO_2, Phys. Rev. B, 81.

123. Yang, et al. (2002). Preparation and Properties of Phenolic Resin-Based Activated Carbon Spheres with Controlled Pore Size Distribution, *Carbon, 40(5)*, 911–916.

124. Liu, et al. (2009). Carbon Nano Tube Based Artificial Water Channel Protein, Membrane Perturbation and Water Transportation, Nano Lett, *9(4)*, 1386–1394.

125. Zhang, et al. (2011). Interfacial Characteristics of Carbon Nano Tube-Polyethylene Composites Using Molecular Dynamics Simulations, ISRN Materials Science, Article ID 145042.

126. Mangun, L. et al. (2001). Surface Chemistry, Pore Sizes and Adsorption Properties of Activated Carbon Fibers and Precursors Treated with Ammonia, *Carbon, 139(11)*, 1809–1820.

127. Irle, et al. (2009). Milestones in Molecular Dynamics Simulations of Single-Walled Carbon Nano Tube Formation, Brief Critical Review, Nano Res, *2(8)*, 755–767.

128. Guo-zhuo, et al. (2009). Regulation of Pore Size Distribution in Coal-Based Activated Carbon, *New Carbon Materials, 24(2)*, article ID: 1007–8827.

129. Zang, et al. (2009). A Comparative Study of Young's modulus of Single-Walled Carbon Nano Tube by CPMD, MD and First Principle Simulations, *Computational Materials Science, 46(4)*, 621–625.

130. Frankland, et al. (2002). Molecular Simulation of the Influence of Chemical Cross-Links on the Shear Strength of Carbon Nano Tube-Polymer Interfaces, *J. Phys. Chem. B, 106(2)*, 3046–3048.

131. Frankland, et al. (2002). Simulation for Separation of Hydrogen and Carbon Monoxide by Adsorption on Single-Walled Carbon Nano Tubes, *Fluid Phase Equilibrium, 194–197(10)*, 297–307.

132. Han, et al. (2007). Molecular Dynamics Simulations of the Elastic Properties of Polymer/Carbon Nano Tube Composites, *Computational Materials Science, 39(8)*, 315–323.

133. Okhovat, et al. (2012). Pore Size Distribution Analysis of Coal-Based Activated Carbons, Investigating the Effects of Activating Agent and Chemical Ratio, ISRN Chemical Engineering.

134. Lehtinen, et al. (2010). Effects of Ion Bombardment on a Two-Dimensional Target, Atomistic Simulations of Graphene Irradiation, Physical Review B, *81*.

135. Shigeo Maruyama (2003). A Molecular Dynamics Simulation of Heat Conduction of a Finite Length Single-Walled Carbon Nano Tube, *Micro Scale Thermo Physical Engineering, 7*, 41–50.

136. Williams, et al. (2000). Monte Carlo simulations of H_2 Physisorption in Finite-Diameter Carbon Nano Tube Ropes, *Chemical Physics Letters, 320(6)*, 352–358.

137. Diao, et al. (2008). Molecular Dynamics Simulations of Carbon Nanotube/Silicon Interfacial Thermal Conductance, *The Journal of Chemical Physics*, 128.
138. Noel, et al. (2010). On the Use of Symmetry in the AB Initio Quantum Mechanical Simulation of Nano Tubes and Related Materials, *J Comput Chem.*, *31*, 855–862.
139. Nicholls, D. et al. (2012). Water Transport Through (7, 7) Carbon Nano Tubes of Different Lengths using Molecular Dynamics, *Micro Fluidics and Nano Fluidics, 1–4*, 257–264.
140. Ribas, A. et al. (2009). Nano Tube Nucleation versus Carbon-Catalyst Adhesion–Probed by Molecular Dynamics Simulations, *J. Chem. Phys,* 131.
141. Sanz-Navarro, F, et al. (2010). Molecular Dynamics Simulations of Metal Clusters Supported on Fishbone Carbon Nano Fibers, *J. Phys. Chem. C*, *114*, 3522–3530.
142. Shibuta, et al. (2003). Bond-Order Potential for Transition Metal Carbide Cluster for the Growth Simulation of a Single-Walled Carbon Nano Tube Department of Materials Engineering, The University of Tokyo.
143. Mao, et al. (1999). Molecular Dynamics Simulations of the Filling and Decorating of Carbon Nano Tubules, *Nanotechnology, 10*, 273–277.
144. Mao, et al. (2010). Molecular Simulation Study of CH_4/H_2 Mixture Separations Using Metal Organic Framework Membranes and Composites, *J. Phys. Chem. C, 114*, 13047–13054.
145. Thomas, A. et al. (2009). Pressure-driven Water Flow through Carbon Nano Tubes, Insights from Molecular Dynamics Simulation, Carnegie Mellon University, USA, Department of Mechanical Engineering.
146. J. S. I. et al. (2009). The Study of Controlling Pore Size on Electro Spun Carbon Nano Fibers for Hydrogen Adsorption, *Journal of Colloid and Interface Science, 318*, 42–49.
147. A. P. T. et al. (2007). How Realistic is the Pore Size Distribution Calculated from Adsorption Isotherms if Activated Carbon is composed of Fullerene-like Fragments, *Physical Chemistry Chemical Physics, 9*, 5919–5927.
148. L. Z. et al. (2013). Controlling the Effective Surface Area and Pore Size Distribution of sp^2 Carbon Materials and Their Impact on the Capacitance Performance of These Materials, *J. Am. Chem. Soc.*, *135*, 5921–5929.
149. A. S. et al. (2011). Reliable Prediction of Pore Size Distribution for Nano-Sized Adsorbents with Minimum Information Requirements, *Chemical Engineering Journal, 171*, 69–80.
150. A. P. T. et al. (2002). What Kind of Pore Size Distribution is assumed in the Dubinin–Astakhov Adsorption Isotherm Equation? *Carbon, 40*, 2879–2886.
151. P. K. et al. (2002). The Application of a CONTIN Package for the Evaluation of Micro Pore Size Distribution Functions, *Langmuir, 18*, 5406–5413.
152. P. A. G. et al. (2004). Estimating the Pore Size Distribution of Activated Carbons from Adsorption Data of Different Adsorbents by Various Methods, *Journal of Colloid and Interface Science, 273*, 39–63.
153. J. J. et al. (2013). 2D-NLDFT Adsorption Models for Carbon Slit-Shaped Pores with Surface Energetically Heterogeneity and Geometrical Corrugation, *Carbon.*
154. E. A. U. et al. (2006). Pore Size Distribution Analysis of Activated Carbons, Application of Density Functional Theory Using Non Graphitized Carbon Black as a Reference System, *Carbon, 44*, 653–663.

155. J. L. et al. (2013). Density Functional Theory Methods for Characterization of Porous Materials, Colloids and Surfaces A, *Physicochemical and Engineering Aspects, 437*, 3–32.

156. P. K. et al. (2008). Adsorption-Induced Deformation of Micro Porous Carbons, Pore Size Distribution Effect, *Langmuir, 24*, 6603–6608.

157. R. L. et al. (2003). Estimation of the Pore-Size Distribution Function from the Nitrogen Adsorption Isotherm, Comparison of Density Functional Theory and the Method of Do and co-workers, *Carbon, 41*, 1113–1125.

158. M. N. S. et al. (2013). Robust PSD Determination of Micro and Meso-Pore Adsorbents via Novel Modified U Curve Method, *Chemical Engineering Research and Design, 9*, 51–62.

159. H. P. (2013). Fullerene-like Models for Micro Porous Carbon, *a Review, Journal of Materials Science, 48(2)*, 565–577.

160. P. K. et al. (2004). Description of Benzene Adsorption in Slit-like Pores. Theoretical Foundations of the Improved Horvath–Kawazoe Method, *Carbon, 42*, 851–864.

161. Mccallun, C. L. et al., (1999). A Molecular Model for Adsorption of Water on Activated Carbon, Comparison of Simulation and Experiment, *Langmuir, 15*, 533–544.

162. M. I. (2009). Pores in Carbon Materials-Importance of their Control, *New Carbon Materials, 24(3)*, 193–232.

163. B. H. et al. (1995). Monte Carlo simulation and Global Optimization without Parameters, *Physical Review Letters, 74(12)*, 2151.

164. A. M. P. (2004). Introduction to Molecular Dynamics Simulation, Computational Soft Matter, *From Synthetic Polymers to Proteins, 23*, 1–28.

165. L. D. P. et al. (2009). A Guide to Monte Carlo simulations in Statistical Physics, Cambridge University Press.

166. A. B. et al. (2013). Molecular Simulations of Supercritical Fluid Permeation through Disordered Micro Porous Carbons, *Langmuir, 29*, 9985–9990.

167. P. K. et al. (2005). Improvement of the Derjaguin-Broekhoff-De Boer Theory for the Capillary Condensation/Evaporation of Nitrogen in Spherical Cavities and its Application for the Pore Size Analysis of Silica with Ordered Cage like Mesopore, *Langmuir, 21*, 10530–10536.

168. M. C. (1983). Micro Canonical Monte Carlo Simulation, the American Physical Society, *50(19)*.

169. P. A. G. et al. (2007). Bimodal Pore Size Distributions for Carbons, Experimental Results and Computational Studies, *Journal of Colloid and Interface Science, 310*, 205–216.

170. A. P. T. et al. (2008). Testing Isotherm Models and Recovering Empirical Relationships for Adsorptions in Micro porous Carbons Using Virtual Carbons Models and Grand Canonical Monte Carlo Simulations, *Journal of Physics Condensed Matter, 20*, 385212–385237.

171. S. K. et al. (2012). Recent Advances in Molecular Dynamics Simulations of Gas Diffusion in Metal Organic Frame Works, Molecular Dynamics-Theoretical Developments and *Applications in Nanotechnology and Energy*, 255–280.

172. F. J. K. et al. (2010). Temperature and Pore Size Effects on Diffusion in Single-Wall Carbon Nanotubes, *Journal of the University of Chemical Technology and Metallurgy, 45(2)*, 161–168.

173. K. A. F. et al. (1991). Theoretical Foundations of Dynamical Monte Carlo Simulations, *Journal of Chemical Physics, 95(2),* 1090.

174. E. E. L. et al. (1984). Monte Carlo Simulation of Markov Unreliability Models, *Nuclear Engineering and Design, 77(1),* 49–62.

175. J. J. P. et al. (2009). Pore Size Distribution and Supercritical Hydrogen Adsorption in Activated Carbon Fibers, *Nanotechnology, 20(20),* 204012.

176. C. L. M. et al. (2001). Surface Chemistry, Pore Sizes and Adsorption Properties of Activated Carbon Fibers and Precursors Treated with Ammonia, *Carbon, 39,* 1809–1820.

177. G. G. Z. et al. (2009). Regulation of Pore Size Distribution in Coal-Based Activated Carbon, *New Carbon Materials, 24(2),* 141–146.

178. K. S. et al. (2012). Bio Composite Fiber of Calcium Alginate/Multi-Walled Carbon Nano Tubes with Enhanced Adsorption Properties for Ionic Dyes, *Carbohydrate Polymers, 90,* 399–406.

179. Li, Y. et al. (2010). Removal of Copper from Aqueous Solution by Carbon Nano Tube/Calcium Alginate Composites, *Journal of Hazardous Materials, 177,* 876–880.

180. Kumar, R. et al. (2013). DBSA Doped Poly Aniline/Multi-Walled Carbon Nano Tubes Composite for High Efficiency Removal of Cr(VI) from Aqueous Solution, *Chemical Engineering Journal, 228,* 748–755.

181. D. S. et al. (2012). Application of Polyaniline and Multi Walled Carbon Nano Tube Magnetic Composites for Removal of Pb(II), *Chemical Engineering Journal, 185, 186,* 144 150.

182. Z. S. et al. (2010). Carbon Nano Tube Polymer Composites, Chemistry, Processing, Mechanical and Electrical Properties, *Progress in Polymer Science, 35,* 357–401.

183. Ch, T. W. et al. (2010). An Assessment of the Science and Technology of Carbon Nano Tube-Based Fibers and Composites, *Composites Science and Technology, 70,* 1–19.

184. J. J. V. et al. (2012). The Hierarchical Structure and Properties of Multi Functional Carbon Nano Tube Fiber Composites, *Carbon, 50,* 1227–1234.

185. M. D. et al. (2012). Growth of Long and Aligned Multi-Walled Carbon Nano Tubes on Carbon and Metal Substrates, *Nanotechnology, 23,* 105604–105612.

186. T. M. et al. (2012). Improved Graphitization and Electrical Conductivity of Suspended Carbon Nano Fibers Derived from Carbon Nano Tubes/Poly acrylonitrile Composite by Directed Electro Spinning, *Carbon, 50,* 1753–1761.

187. K. N. et al. (2010). Enhancing the Thermal Conductivity of Polyacrylonitrile and Pitch-Based Carbon Fibers by Grafting Carbon Nano Tubes on them, *Carbon, 48,* 1849–1857.

188. G. X. et al. (2011). Binder-Free Activated Carbon/Carbon Nanotube Paper Electrodes for Use in Super capacitors, *Nano Research, 4(9),* 870–881.

189. B. D. et al. (2012). Schiff Base-Chitosan Grafted Multi Walled Carbon Nano Tubes as a Novel Solid-Phase Extraction Adsorbent for Determination of Heavy Metal by ICP-MS, *Journal of Hazardous Materials, 103–110,* 219, 220.

190. C. T. H. et al. (2009). Synthesis of Carbon Nanotubes on Carbon Fabric for Use as Electrochemical Capacitor, *Micro porous and Mesoporous Materials, 122,* 155–159.

191. H. G. et al. (2013). Removal of Anionic Azo Dyes from Aqueous Solution using Magnetic Polymer Multi-Wall Carbon Nanotube Nanocomposite as adsorbent, *Chemical Engineering Journal, 223,* 84–90.
192. Zhu, M., & Diao, G. (2011). Review on the Progress in Synthesis and Application of Magnetic Carbon Nano composites, *Nanoscale, 3,* 2748–2767.

APPENDIX

Program 1: Test a Random Number Generator in Monte Carlo Simulation.
Note, as an exercise the student may wish to insert other random number-generators or add tests to this simple program.

```
c**********************************************************
****
c This program is used to perform a few very simple tests of a random
c number generator. A congruential generator is being tested
c**********************************************************
****
      Real*8 Rnum(100000),Rave,R2Ave,Correl,SDev
      Integer Iseed,num
      open(Unit=1,®le='result_testrng_02')
      PMod = 2147483647.0D0
      DMax = 1.0D0/PMod
c*******
c Input
c*******
      write(*,800)
800  format('enter the random number generator seed ')
      read(*,921) Iseed
921  format(i5)
      write(*,801) Iseed
      write(1,801) Iseed
801  format('The random number seed is', I8)
      write(*,802)
802  format('enter the number of random numbers to be generated')
      read(*,921) num
      write(*,803) num
      write(1,803)num
803   format ('number of random numbers to be generated = ',i8)
c***************************
c Initialize variables, vectors
```

```fortran
c****************************
do 1 i=2,10000
1   Rnum(i)=0.0D0
    Ravc=0.D0
    Correl=0.0D0
    R2Ave=0.0D0
    SDev=0.0D0
c************************
c Calculate random numbers
c************************
    Rnum(1)=Iseed*DMax
    Write(*,931) Rnum(1),Iseed
    Do 10 i=2,num
    Rnum(i)=cong16807(Iseed)
    if (num.le.100) write(*,931) Rnum(i),Iseed
931  format(f10.5,i15)
10    continue
    Rave=Rnum(1)
    R2Ave=Rnum(1)**2
    Do 20 i=2,num
    Correl=Correl+Rnum(i)*Rnum(i-1)
    Rave=Rave+Rnum(i)
20    R2Ave=R2Ave+Rnum(i)**2
    Rave=Rave/num
    SDev=Sqrt((R2Ave/num-Rave**2)/(num-1))
    Correl=Correl/(num-1)-Rave*RAve
c*******
c Output
c*******
    write(*,932) Rave,SDev,Correl
932  format('Ave. random number = ',F10.6, ' +/-', F10.6,
1 / ' ''nn''-correlation = ' F10.6)
    write(1,932) Rave,SDev,Correl
999  format(f12.8)
    close (1)
    stop
    end
    FUNCTION Cong16807(ISeed)
```

```
c*****************************************************
c This is a simple congruential random number generator
c*****************************************************
      INTEGER ISeed,IMod
      REAL*8 RMod,PMod,DMax
      RMod = DBLE(ISeed)
      PMod = 2147483647.0D0
      DMax = 1.0D0/PMod
      RMod = RMod*16807.0D0
      IMod = RMod*DMax
      RMod = RMod – PMod*IMod
      cong16807=rmod*DMax
      Iseed=Rmod
      RETURN
      END
```

Program 2: A Good Routine For Generating a Table of Random Numbers in Monte Carlo Simulation.

```
C*****************************************************************
******
C This program uses the R250/R521 combined generator described in:
C A. Heuer, B. Duenweg and A.M. Ferrenberg, Comp. Phys. Comm. 103,
1
C 1997). It generates a vector, RanVec, of length RanSize 31-bit random
C integers. Multiply by RMaxI to get normalized random numbers. You
C will need to test whether RanCnt will exceed RanSize. If so, call
C GenRan again to generate a new block of RanSize numbers. Always
C remember to increment RanCnt when you use a number from the table.
C*****************************************************************
******
      IMPLICIT NONE
      INTEGER RanSize,Seed,I,RanCnt,RanMax
      PARAMETER(RanSize = 10000)
      PARAMETER(RanMax = 2147483647)
      INTEGER RanVec(RanSize),Z1(250+RanSize),Z2(521+RanSize)
      REAL*8 RMaxI
      PARAMETER (RMaxI = 1.0D0/(1.0D0*RanMax))
      COMMON/MyRan/RanVec,Z1,Z2,RanCnt
```

```
      SAVE
      Seed = 432987111
C****************************************
C Initialize the random number generator.
C****************************************
      CALL InitRan(Seed)*
C*********************************************************
******

C If the 10 numbers we need pushes us past the end of the RanVec vector,
C call GenRan. Since we just called InitRan, RanCnt = RanSize we must
C call it here.
C*********************************************************
******
      IF ((RanCnt + 10) .GT. RanSize) THEN
C** Generate RanSize numbers and reset the RanCnt counter to 1
      Call GenRan
      END IF
      Do I = 1,10
      WRITE(*,*) RanVec(RanCnt + I – 1),RMaxI*RanVec(RanCnt + I – 1)
      End Do
      RanCnt = RanCnt + 10
C*********************************************************
******

C Check to see if the 10 numbers we need will push us past the end
C of the RanVec vector. If so, call GenRan.
C*********************************************************
******
      IF ((RanCnt + 10) .GT. RanSize) THEN
C** Generate RanSize numbers and reset the RanCnt counter to 1
      Call GenRan
      END IF
      Do I = 1,10
      WRITE(*,*) RanVec(RanCnt + I – 1),RMaxI*RanVec(RanCnt + I – 1)
      End Do
      RanCnt = RanCnt + 10
      END
      SUBROUTINE InitRan(Seed)
```

```
C*****************************************************
******

C Initialize the R250 and R521 generators using a congruential generator
C to set the individual bits in the 250/521 numbers in the table. The
C R250 and R521 are then warmed-up by generating 1000 numbers.
C*****************************************************
******
      IMPLICIT NONE
      REAL*8 RMaxI,RMod,PMod
      INTEGER RanMax,RanSize
      PARAMETER(RanMax = 2147483647)
      PARAMETER(RanSize = 100000)
      PARAMETER (RMaxI = 1.0D0/(1.0D0*RanMax))
      INTEGER Seed,I,J,K,IMod,IBit
      INTEGER RanVec(RanSize),Z1(250+RanSize),Z2(521+RanSize)
      INTEGER RanCnt
      COMMON/MyRan/RanVec,Z1,Z2,RanCnt
      SAVE
      RMod = DBLE(Seed)
      PMod = DBLE(RanMax)
C***********************************
C Warm up a congruential generator
C***********************************
      Do I = 1,1000
      RMod = RMod*16807.0D0
      IMod = RMod/PMod
      RMod = RMod – PMod*IMod
      End Do
C*****************************************************
******

C Now ®ll up the tables for the R250 & R521 generators: This
C requires random integers in the range 0±> 2*31 1. Iterate a
C strange number of times to improve randomness.
C*****************************************************
******
      Do I = 1,250
      Z1(I) = 0
      IBit = 1
```

```
      Do J = 0,30
      Do K = 1,37
      RMod = RMod*16807.0D0
      IMod = RMod/PMod
      RMod = RMod – PMod*IMod
      End Do
C** Now use this random number to set bit J of X(I).
      IF (RMod .GT. 0.5D0*PMod) Z1(I) = IEOR(Z1(I),IBit)
      IBit = IBit*2
End Do
End Do
Do I = 1,521
      Z2(I) = 0
      IBit = 1
      Do J = 0,30
      Do K = 1,37
      RMod = RMod*16807.0D0
      IMod = RMod/PMod
      RMod = RMod – PMod*IMod
      End Do
C** Now use this random number to set bit J of X(I).
      IF (RMod .GT. 0.5D0*PMod) Z2(I) = IEOR(Z2(I),IBit)
      IBit = IBit*2
      End Do
      End Do
C*********************************************************
******
C Perform a few iterations of the R250 and R521 random number genera-
tors
C to eliminate any effects due to 'poor' initialization.
C*********************************************************
******
      Do I = 1,1000
      Z1(I+250) = IEOR(Z1(I),Z1(I+147))
      Z2(I+521) = IEOR(Z2(I),Z2(I+353))
      End Do
      Do I = 1,250
      Z1(I) = Z1(I + 1000)
```

```
End Do
Do I = 1,521
Z2(I) = Z2(I + 1000)
End Do
```
C***
******

C Set the random number counter to RanSize so that a proper checking
C code will force a call to GenRan in the main program.
C***
******

```
RanCnt = RanSize
RETURN
END
SUBROUTINE GenRan
```
C***
******

C Generate vector RanVec (length RanSize) of pseudo-random 31-bit
C integers.
C***
******

```
IMPLICIT NONE
INTEGER RanSize,RanCnt,I
PARAMETER(RanSize = 100000)
INTEGER RanVec(RanSize),Z1(250+RanSize),Z2(521+RanSize)
COMMON/MyRan/RanVec,Z1,Z2,RanCnt
SAVE
```
C***
******

C Generate RanSize pseudo-random numbers using the individual generators
C***
******

```
Do I = 1,RanSize
Z1(I+250) = IEOR(Z1(I),Z1(I+147))
Z2(I+521) = IEOR(Z2(I),Z2(I+353))
End Do
```
C***
******

C Combine the R250 and R521 numbers and put the result into RanVec
C***
******

```
    Do I = 1,RanSize
    RanVec(I) = IEOR(Z1(I+250),Z2(I+521))
    End Do
```
C***
******

C Copy the last 250 numbers generated by R250 and the last 521 numbers
C from R521 into the working vectors (Z1), (Z2) for the next pass.
C***
******

```
    Do I = 1,250
    Z1(I) = Z1(I + RanSize)
    End Do
    Do I = 1,521
    Z2(I) = Z2(I + RanSize)
    End Do
```
C***

C Reset the random number counter to 1.
C***

```
    RanCnt = 1
    RETURN
    END
```

INDEX

9 781774 633656